An Introduction to LINEAR ALGEBRA AND TENSORS

M. A. AKIVIS

V. V. GOLDBERG

Revised English Edition
Translated and Edited by
Richard A. Silverman

DOVER PUBLICATIONS, INC.

NEW YORK

Published in Canada by General Publishing Company, Ltd., 30 Lesmill Road, Don Mills, Toronto, Ontario.
Published in the United Kingdom by Constable and Company, Ltd., 10 Orange Street, London WC2H 7EG.

This Dover edition, first published in 1977, is an unabridged and unaltered republication of the revised English edition published by Prentice-Hall, Inc., Englewood Cliffs, N. J., in 1972 under the title *Introductory Linear Algebra.*

International Standard Book Number: 0-486-63545-7
Library of Congress Catalog Card Number: 77-78589

Manufactured in the United States of America
Dover Publications, Inc.
180 Varick Street
New York, N. Y. 10014

CONTENTS

Editor's Preface, vii

1 LINEAR SPACES, Page 1.

1. Basic Concepts, 1.
2. Linear Dependence, 4.
3. Dimension and Bases, 8.
4. Orthonormal Bases. The Scalar Product, 12.
5. The Vector Product. Triple Products, 16.
6. Basis Transformations. Tensor Calculus, 22.
7. Topics in Analytic Geometry, 29.

2 MULTILINEAR FORMS AND TENSORS, Page 38.

8. Linear Forms, 38.
9. Bilinear Forms, 41.
10. Multilinear Forms. General Definition of a Tensor, 44.
11. Algebraic Operations on Tensors, 50.
12. Symmetric and Antisymmetric Tensors, 55.

3 LINEAR TRANSFORMATIONS, Page 64.

13. Basic Concepts, 64.
14. The Matrix of a Linear Transformation and Its Determinant, 68.
15. Linear Transformations and Bilinear Forms, 78.
16. Multiplication of Linear Transformations and Matrices, 87.
17. Inverse Transformations and Matrices, 94.
18. The Group of Linear Transformations and Its Subgroups, 98.

4 FURTHER TOPICS, Page 107.

19. Eigenvectors and Eigenvalues, 107.
20. The Case of Distinct Eigenvalues, 117.
21. Matrix Polynomials and the Hamilton-Cayley Theorem, 121.
22. Eigenvectors of a Symmetric Transformation, 124.
23. Diagonalization of a Symmetric Transformation, 127.
24. Reduction of a Quadratic Form to Canonical Form, 133.
25. Representation of a Nonsingular Transformation, 138.

SELECTED HINTS AND ANSWERS, Page 144.

BIBLIOGRAPHY, Page 161.

INDEX, Page 163.

EDITOR'S PREFACE

The present book, stemming from the first four chapters of the authors' *Tensor Calculus* (Moscow, 1969), constitutes a lucid and completely elementary introduction to linear algebra. The treatment is virtually self-contained. In fact, the mathematical background assumed on the part of the reader hardly exceeds a smattering of calculus and a casual acquaintance with determinants. A special merit of the book, reflecting its lineage, is its free use of tensor notation, in particular the Einstein summation convention. Each of the 25 sections is equipped with a problem set, leading to a total of over 250 problems. Hints and answers to most of these problems can be found at the end of the book.

As usual, I have felt free to introduce a number of pedagogical and mathematical improvements that occurred to me in the course of the translation.

R. A. S.

1

LINEAR SPACES

1. Basic Concepts

In studying analytic geometry, the reader has undoubtedly already encountered the concept of a *free vector*, i.e., a directed line segment which can be shifted in space parallel to its original direction. Such vectors are usually denoted by boldface Roman letters like $\mathbf{a}, \mathbf{b}, \ldots, \mathbf{x}, \mathbf{y}, \ldots$ It can be assumed for simplicity that the vectors all have the same initial point, which we denote by the letter O and call the *origin of coordinates.*

Two operations on vectors are defined in analytic geometry:

a) Any two vectors $\mathbf{x}$ and $\mathbf{y}$ can be added (in that order), giving the *sum* $\mathbf{x} + \mathbf{y}$;

b) Any vector $\mathbf{x}$ and (real) number α can be multiplied, giving the *product* $\lambda \cdot \mathbf{x}$ or simply $\lambda\mathbf{x}$.

The set of all spatial vectors is *closed* with respect to these two operations, in the sense that the sum of two vectors and the product of a vector with a number are themselves both vectors.

The operations of addition of vectors $\mathbf{x}, \mathbf{y}, \mathbf{z}, \ldots$ and multiplication of vectors by real numbers $\lambda, \mu, \ldots$ have the following properties:

1) $\mathbf{x} + \mathbf{y} = \mathbf{y} + \mathbf{x}$;
2) $(\mathbf{x} + \mathbf{y}) + \mathbf{z} = \mathbf{x} + (\mathbf{y} + \mathbf{z})$;
3) There exists a *zero vector* $\mathbf{0}$ such that $\mathbf{x} + \mathbf{0} = \mathbf{x}$;
4) Every vector $\mathbf{x}$ has a *negative* (*vector*) $\mathbf{y} = -\mathbf{x}$ such that $\mathbf{x} + \mathbf{y} = \mathbf{0}$;
5) $1 \cdot \mathbf{x} = \mathbf{x}$;
6) $\lambda(\mu\mathbf{x}) = (\lambda\mu)\mathbf{x}$;

7) $(\lambda + \mu)\mathbf{x} = \lambda\mathbf{x} + \mu\mathbf{x}$;
8) $\lambda(\mathbf{x} + \mathbf{y}) = \lambda\mathbf{x} + \lambda\mathbf{y}$.

However, operations of addition and multiplication by numbers can be defined for sets of elements other than the set of spatial vectors, such that the sets are closed with respect to the operations and the operations satisfy the properties 1)–8) just listed. Any such set of elements is called a *linear space* (or *vector space*), conventionally denoted by the letter L. The elements of a vector space L are often called *vectors*, by analogy with the case of ordinary vectors.

Example 1. The set of all vectors lying on a given straight line l forms a linear space, since the sum of two such vectors and the product of such a vector with a real number is again a vector lying on l, while properties 1)–8) are easily verified. This linear space will be denoted by L_1.†

Example 2. The set of all vectors lying in a given plane is also closed with respect to addition and multiplication by real numbers, and clearly satisfies properties 1)–8). Hence this set is again a linear space, which we denote by L_2.

Example 3. Of course, the set of all spatial vectors is also a linear space, denoted by L_3.

Example 4. The set of all vectors lying in the xy-plane whose initial points coincide with the origin of coordinates and whose end points lie in the first quadrant is not a linear space, since it is not closed with respect to multiplication by real numbers. In fact, the vector $\lambda\mathbf{x}$ does not belong to the first quadrant if $\lambda < 0$.

Example 5. Let L_n be the set of all ordered n-tuples

$$\mathbf{x} = (x_1, x_2, \ldots, x_n), \qquad \mathbf{y} = (y_1, y_2, \ldots, y_n), \ldots$$

of real numbers $x_1, \ldots, y_n, \ldots$ with addition of elements and multiplication of an element by a real number λ defined by

$$\begin{aligned} \mathbf{x} + \mathbf{y} &= (x_1 + y_1, x_2 + y_2, \ldots, x_n + y_n), \\ \lambda\mathbf{x} &= (\lambda x_1, \lambda x_2, \ldots, \lambda x_n). \end{aligned} \tag{1}$$

Then L_n is a linear space, since L_n is closed with respect to the operations (1) which are easily seen to satisfy properties 1)–8). For example, the zero element in L_n is the vector

$$\mathbf{0} = (0, 0, \ldots, 0),$$

while the negative of the vector $\mathbf{x}$ is just

$$-\mathbf{x} = (-x_1, -x_2, \ldots, -x_n).$$

Example 6. As is easily verified, the set of all polynomials

† Concerning the meaning of the subscript here and below, see Sec. 3.

$$P(t) = a_0 + a_1 t + \ldots + a_n t^n$$

of degree not exceeding n is a linear space, with addition and multiplication by real numbers defined in the usual way.

Example 7. The set of all functions $\varphi(t)$ continuous in an interval $[a, b]$ is also a linear space (with the usual definition of addition and multiplication by real numbers). We will denote this space by $C[a, b]$.

PROBLEMS

1. Which of the following are linear spaces:

a) The set of all vectors† of the space L_2 (recall Example 2) with the exception of vectors parallel to a given line;

b) The set of all vectors of the space L_2 whose end points lie on a given line;

c) The set of all vectors of the space L_3 (recall Example 3) whose end points do not belong to a given line?

2. Which of the following sets of vectors $\mathbf{x} = (x_1, x_2, \ldots, x_n)$ in the space L_n (recall Example 5) are linear spaces:

a) The set such that $x_1 + x_2 + \cdots + x_n = 0$;

b) The set such that $x_1 + x_2 + \cdots + x_n = 1$;

c) The set such that $x_1 = x_3$;

d) The set such that $x_2 = x_4 = \cdots$;

e) The set such that x_1 is an integer;

f) The set such that x_1 or x_2 vanishes?

3. Does the set of all polynomials of degree n (cf. Example 6) form a linear space?

4. Let R^+ denote the set of *positive* real numbers. Define the "sum" of two numbers $p \in R^+$, $q \in R^+$‡ as pq and the "product" of a number $p \in R^+$ with an arbitrary real number λ as p^λ. Is R^+ a linear space when equipped with these operations? What is the "zero element" in R^+? What is the "negative" of an element $p \in R^+$?

5. Prove that the set of solutions of the homogeneous linear differential equation

$$y^{(n)} + p_1(x)y^{(n-1)} + \cdots + p_{n-1}(x)y' + p_n(x)y = 0$$

of order n forms a linear space.

6. Let L' be a nonempty subset of a linear space L, i.e., a subset of L containing at least one vector. Then L' is said to be a *linear subspace* of L if L' is itself a linear space with respect to the operations (of addition and multiplication by numbers) already introduced in L, i.e., if $\mathbf{x} + \mathbf{y} \in L'$, $\lambda\mathbf{x} \in L'$ whenever $\mathbf{x} \in L'$, $\mathbf{y} \in L'$. The simplest subspaces of every linear space L (the *trivial* subspaces) are the space L itself and the space $\{\mathbf{0}\}$ consisting of the single element $\mathbf{0}$ (the

---

† As agreed at the beginning of the section, we regard all vectors as emanating from the origin of coordinates.

‡ As usual, the symbol $\in$ means "is an element of" or "belongs to."

zero element). By the *sum* of two linear subspaces L' and L'' of a linear space is meant the set, denoted by $L' + L''$, of all vectors in L which can be represented in the form $\mathbf{x} = \mathbf{x}' + \mathbf{x}''$ where $\mathbf{x}' \in L'$, $\mathbf{x}'' \in L''$. By the *intersection* of two linear subspaces L' and L'' of a linear space L is meant the set, denoted by $L' \cap L''$, of all vectors in L which belong to both L' and L''.

Prove that the sum and intersection of two linear subspaces of a linear space L are themselves linear subspaces of L.

7. Describe all linear subspaces of the space L_3.

8. Which sets of vectors in Prob. 2 are linear subspaces of the space L_n?

2. Linear Dependence

2.1. Let $\mathbf{a}, \mathbf{b}, \ldots, \mathbf{e}$ be vectors of a linear space L, and let $\alpha, \beta, \ldots, \epsilon$ be real numbers. Then the vector

$$\mathbf{x} = \alpha\mathbf{a} + \beta\mathbf{b} + \cdots + \epsilon\mathbf{e}$$

is called a *linear combination* of the vectors $\mathbf{a}, \mathbf{b}, \ldots, \mathbf{e}$, and the numbers $\alpha, \beta, \ldots, \epsilon$ are called the *coefficients* of the linear combination.

If $\alpha = \beta = \cdots = \epsilon = 0$, then obviously $\mathbf{x} = \mathbf{0}$. But there may also exist a linear combination of the vectors $\mathbf{a}, \mathbf{b}, \ldots, \mathbf{e}$ which equals zero even though the coefficients $\alpha, \beta, \ldots, \epsilon$ are not all zero; in this case, the vectors $\mathbf{a}, \mathbf{b}, \ldots, \mathbf{e}$ are said to be *linearly dependent*. In other words, the vectors $\mathbf{a}, \mathbf{b}, \ldots, \mathbf{e}$ are linearly dependent if and only if there are real numbers $\alpha, \beta, \ldots, \epsilon$ not all equal to zero such that

$$\alpha\mathbf{a} + \beta\mathbf{b} + \cdots + \epsilon\mathbf{e} = \mathbf{0}. \tag{1}$$

Suppose (1) holds if and only if the numbers $\alpha, \beta, \ldots, \epsilon$ are all zero. Then $\mathbf{a}, \mathbf{b}, \ldots, \mathbf{e}$ are said to be *linearly independent*.

We now prove some simple properties of linearly dependent vectors.

THEOREM 1. *If the vectors* $\mathbf{a}, \mathbf{b}, \ldots, \mathbf{e}$ *are linearly dependent, then one of the vectors can be represented as a linear combination of the others. Conversely, if one of the vectors* $\mathbf{a}, \mathbf{b}, \ldots, \mathbf{e}$ *is a linear combination of the others, then the vectors are linearly dependent.*

Proof. If the vectors $\mathbf{a}, \mathbf{b}, \ldots, \mathbf{e}$ are linearly dependent, then

$$\alpha\mathbf{a} + \beta\mathbf{b} + \cdots + \epsilon\mathbf{e} = \mathbf{0},$$

where the coefficients $\alpha, \beta, \ldots, \epsilon$ are not all zero. Suppose, for example, that $\alpha \neq 0$. Then

$$\mathbf{a} = -\frac{\beta}{\alpha}\mathbf{b} - \cdots - \frac{\epsilon}{\alpha}\mathbf{e},$$

which proves the first assertion.

Conversely, if one of the vectors $\mathbf{a}, \mathbf{b}, \ldots, \mathbf{e}$, say $\mathbf{a}$, is a linear combination of the others, then

$$\mathbf{a} = m\mathbf{b} + \cdots + p\mathbf{e},$$

and hence

$$1 \cdot \mathbf{a} + (-m)\mathbf{b} + \cdots + (-p)\mathbf{e} = \mathbf{0},$$

i.e., the vectors $\mathbf{a}, \mathbf{b}, \ldots, \mathbf{e}$ are linearly dependent. ∎†

THEOREM 2. *If some of the vectors* $\mathbf{a}, \mathbf{b}, \ldots, \mathbf{e}$ *are linearly dependent, then so is the whole system.*

Proof. Suppose, for example, that $\mathbf{a}$ and $\mathbf{b}$ are linearly dependent. Then

$$\alpha\mathbf{a} + \beta\mathbf{b} = \mathbf{0},$$

where at least one of the coefficients α and β is nonzero. But then

$$\alpha\mathbf{a} + \beta\mathbf{b} + 0 \cdot \mathbf{c} + \cdots + 0 \cdot \mathbf{e} = \mathbf{0},$$

where at least one of the coefficients of the linear combination on the left is nonzero, i.e., the whole system of vectors $\mathbf{a}, \mathbf{b}, \ldots, \mathbf{e}$ is linearly dependent. ∎

THEOREM 3. *If at least one of the vectors* $\mathbf{a}, \mathbf{b}, \ldots, \mathbf{e}$ *is zero, then the vectors are linearly dependent.*

Proof. Suppose, for example, that $\mathbf{a} = \mathbf{0}$. Then

$$\alpha\mathbf{a} + 0 \cdot \mathbf{b} + \cdots + 0 \cdot \mathbf{e} = \mathbf{0}$$

for any nonzero number α. ∎

2.2. Next we give some examples of linearly dependent and linearly independent vectors in the space L_3.

Example 1. The zero vector $\mathbf{0}$ is linearly dependent (in a trivial sense), since $\alpha\mathbf{0} = \mathbf{0}$ for any $\alpha \neq 0$. This also follows from Theorem 3.

Example 2. Any vector $\mathbf{a} \neq \mathbf{0}$ is linearly independent, since $\alpha\mathbf{a} = \mathbf{0}$ only if $\alpha = 0$.

Example 3. Two collinear vectors‡ $\mathbf{a}$ and $\mathbf{b}$ are linearly dependent. In fact, if $\mathbf{a} \neq \mathbf{0}$, then $\mathbf{b} = \alpha\mathbf{a}$ or equivalently

$$\alpha\mathbf{a} + (-1)\mathbf{b} = \mathbf{0},$$

while if $\mathbf{a} = \mathbf{0}$, then $\mathbf{a}$ and $\mathbf{b}$ are linearly dependent by Theorem 3.

Example 4. Two noncollinear vectors are linearly independent. In fact, suppose to the contrary that $\alpha\mathbf{a} + \beta\mathbf{b} = \mathbf{0}$ where $\beta \neq 0$. Then

---

† The symbol ∎ stands for Q.E.D. and indicates the end of a proof.

‡ Two or more vectors are said to be *collinear* if they lie on the same line and *coplanar* if they lie in the same plane.

$$\mathbf{b} = -\frac{\alpha}{\beta}\mathbf{a},$$

which implies that **a** and **b** are collinear. Contradiction!

Example 5. Three coplanar vectors are linearly dependent. In fact, suppose the vectors **a**, **b** and **c** are coplanar, while **a** and **b** are noncollinear. Then **c** can be represented as a linear combination

$$\mathbf{c} = \overrightarrow{OC} = \overrightarrow{OA} + \overrightarrow{OB} = \alpha\mathbf{a} + \beta\mathbf{b}$$

(see Figure 1), and hence **a**, **b** and **c** are linearly dependent by Theorem 1. If, on the other hand, the vectors **a** and **b** are collinear, then they are linearly dependent by Example 3, and hence the vectors **a**, **b** and **c** are linearly dependent by Theorem 2.

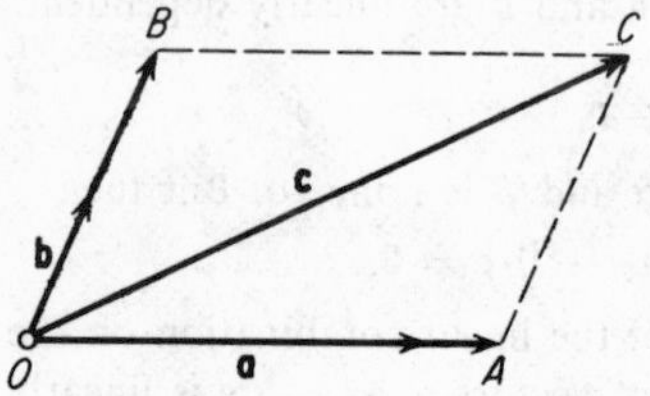

FIGURE 1

Example 6. Three noncoplanar vectors are always linearly independent. The proof is virtually the same as in Example 4 (give the details).

Example 7. Any four spatial vectors are linearly dependent. In fact, if any three vectors are linearly dependent, then all four vectors are linearly dependent by Theorem 2. On the other hand, if there are three linearly independent vectors **a**, **b** and **c** (say), then any other vector **d** can be represented as a linear combination

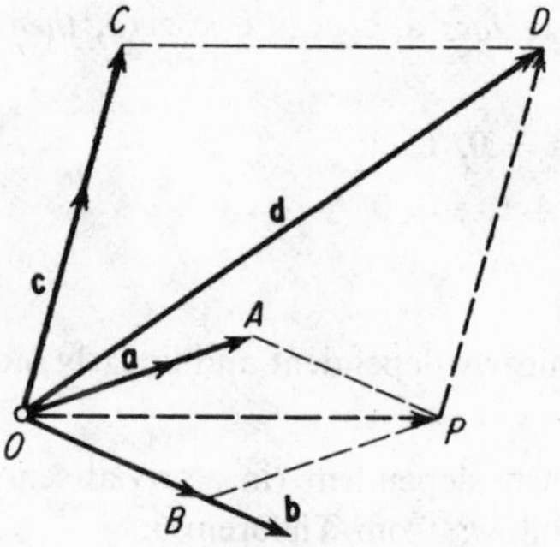

FIGURE 2

$$\mathbf{d} = \overrightarrow{OD} = \overrightarrow{OP} + \overrightarrow{PD} = \overrightarrow{OA} + \overrightarrow{OB} + \overrightarrow{OC} = \lambda\mathbf{a} + \mu\mathbf{b} + \gamma\mathbf{c}$$

(see Figure 2), and hence **a**, **b**, **c** and **d** are linearly dependent by Theorem 1.

Example 8. The vectors

$$\mathbf{e}_1 = (1, 0, \ldots, 0), \quad \mathbf{e}_2 = (0, 1, \ldots, 0), \ldots, \quad \mathbf{e}_n = (0, 0, \ldots, 1)$$

are linearly independent in the space L_n. In fact, the linear combination

$$\alpha_1\mathbf{e}_1 + \alpha_2\mathbf{e}_2 + \ldots + \alpha_n\mathbf{e}_n = (\alpha_1, \alpha_2, \ldots, \alpha_n)$$

equals zero if and only if $\alpha_1 = \alpha_2 = \cdots = \alpha_n = 0$. Let $\mathbf{x} = (x_1, x_2, \ldots, x_n)$ be an arbitrary vector of L_n. Then the system of vectors $\mathbf{e}_1, \mathbf{e}_2, \ldots, \mathbf{e}_n, \mathbf{x}$ is linearly dependent, since $\mathbf{x}$ can be represented in the form

$$\mathbf{x} = x_1\mathbf{e}_1 + x_2\mathbf{e}_2 + \ldots + x_n\mathbf{e}_n.$$

PROBLEMS

1. Let **a** and **b** be linearly independent vectors in L_2. Find the value of α making each of the following pairs of vectors linearly dependent (collinear):

a) $\alpha\mathbf{a} + 2\mathbf{b}, \mathbf{a} - \mathbf{b}$; b) $(\alpha + 1)\mathbf{a} + \mathbf{b}, 2\mathbf{b}$; c) $\alpha\mathbf{a} + \mathbf{b}, \mathbf{a} + \alpha\mathbf{b}$.

Find values of α and β such that

d) $3\mathbf{a} + 5\mathbf{b} = \alpha\mathbf{a} + (2\beta + 1)\mathbf{b}$; e) $(2\alpha - \beta - 1)\mathbf{a} - (3\alpha + \beta + 10)\mathbf{b} = \mathbf{0}$.

2. Let **a**, **b** and **c** be three linearly independent vectors in L_3.

a) For what value of α are the vectors

$$\mathbf{x} = \alpha\mathbf{a} + 4\mathbf{b} + 2\mathbf{c}, \qquad \mathbf{y} = \mathbf{a} + \alpha\mathbf{b} - \mathbf{c}$$

linearly dependent (collinear)?

b) For what value of α are the vectors

$$\mathbf{x} = \alpha\mathbf{a} + \mathbf{b} + 3\mathbf{c}, \qquad \mathbf{y} = \alpha\mathbf{a} - 2\mathbf{b} + \mathbf{c}, \qquad \mathbf{z} = \mathbf{a} - \mathbf{b} + \mathbf{c}$$

linearly dependent (coplanar)?

3. Prove that the following sets of functions are linearly dependent in the space $C[a, b]$ introduced in Sec. 1, Example 7:

a) $\varphi_1(t) = \sin^2 t, \varphi_2(t) = \cos^2 t, \varphi_3(t) = 1$;

b) $\varphi_1(t) = \sin^2 t, \varphi_2(t) = \cos^2 t, \varphi_3(t) = t, \varphi_4(t) = 3, \varphi_5(t) = e^t$;

c) $\varphi_1(t) = \sqrt{t}, \varphi_2(t) = \frac{1}{t^2}, \varphi_3(t) = 0, \varphi_4(t) = t^5$.

4. Prove that the functions

$$\varphi_1(t) = \begin{cases} 0 & \text{if } 0 \leq t < 1, \\ (t-1)^4 & \text{if } 1 \leq t \leq 2, \end{cases}$$

$$\varphi_2(t) = \begin{cases} (t-1)^4 & \text{if } 0 \leq t < 1, \\ 0 & \text{if } 1 \leq t \leq 2 \end{cases}$$

are linearly independent in the space $C[0, 2]$.

5. Prove that the polynomials

$$P_0(t) = 1, P_1(t) = t, \ldots, P_n(t) = t^n$$

are linearly independent in the space of all polynomials of degree not exceeding n.

6. Prove that the space $C[a, b]$ contains an arbitrarily large number of linearly independent vectors.

7. Prove that the vectors

$$\mathbf{a}_1 = (1, 1, 1), \qquad \mathbf{a}_2 = (1, 1, 2), \qquad \mathbf{a}_3 = (1, 2, 3)$$

are linearly dependent in the space L_3.

8. Prove that a set of vectors is linearly dependent if it contains

a) Two equal vectors;

b) Two collinear vectors.

9. Prove that if the vectors $\mathbf{a}_1, \mathbf{a}_2, \mathbf{a}_3$ are linearly independent, then so are the vectors $\mathbf{a}_1 + \mathbf{a}_2, \mathbf{a}_2 + \mathbf{a}_3, \mathbf{a}_3 + \mathbf{a}_1$.

3. Dimension and Bases

The largest number of linearly independent vectors in a linear space L is called the *dimension* of L.

Example 1. There is only one linearly independent vector on a line, any two vectors on the line being linearly dependent. Hence the line is a one-dimensional linear space, which we have already denoted by L_1. Thus the subscript 1 is just the dimension of the space.

Example 2. There are two linearly independent vectors in the plane, but any three vectors in the plane are linearly dependent. Therefore the plane is a two-dimensional space and is accordingly denoted by L_2.

Example 3. There are three linearly independent vectors in space, but any four vectors are linearly dependent. Thus ordinary space is three-dimensional and is denoted by L_3.

Example 4. In Sec. 2, Example 8, we found n linearly independent vectors $e_1, e_2, \ldots, e_n$ in the space whose elements are the vectors $\mathbf{x} = (x_1, x_2, \ldots, x_n)$. On the other hand, it can be shown† that any $n + 1$ vectors in this space are linearly dependent. Therefore this space is n-dimensional and is denoted by L_n.

THEOREM. *Let* $\mathbf{e}_1, \mathbf{e}_2, \ldots, \mathbf{e}_n$ *be any n linearly independent vectors in an n-dimensional linear space L, and let* $\mathbf{x}$ *be any vector of L. Then* $\mathbf{x}$ *has a unique representation as a linear combination of* $\mathbf{e}_1, \mathbf{e}_2, \ldots, \mathbf{e}_n$.

Proof. The vectors $\mathbf{x}, \mathbf{e}_1, \mathbf{e}_2, \ldots, \mathbf{e}_n$ are linearly dependent, since there are more than n of them, i.e., more than the dimension of the space L. Hence there are numbers $\alpha, \alpha_1, \ldots, \alpha_n$ such that

$$\alpha\mathbf{x} + \alpha_1\mathbf{e}_1 + \alpha_2\mathbf{e}_2 + \cdots + \alpha_n\mathbf{e}_n = \mathbf{0},$$

where $\alpha \neq 0$ since otherwise the vectors $\mathbf{e}_1, \mathbf{e}_2, \ldots, \mathbf{e}_n$ would be linearly dependent. Therefore we can represent $\mathbf{x}$ as the following linear combination of $\mathbf{e}_1, \mathbf{e}_2, \ldots, \mathbf{e}_n$:

$$\mathbf{x} = -\frac{\alpha_1}{\alpha}\mathbf{e}_1 - \frac{\alpha_2}{\alpha}\mathbf{e}_2 - \cdots - \frac{\alpha_n}{\alpha}\mathbf{e}_n.$$

Equivalently, we can write

$$\mathbf{x} = x_1\mathbf{e}_1 + x_2\mathbf{e}_2 + \cdots + x_n\mathbf{e}_n \qquad (1)$$

where

$$x_1 = -\frac{\alpha_1}{\alpha}, \quad x_2 = -\frac{\alpha_2}{\alpha}, \ldots, \quad x_n = -\frac{\alpha_n}{\alpha}.$$

---

† See e.g., G. E. Shilov, *Linear Algebra*, translated by R. A. Silverman, Prentice-Hall, Inc., Englewood Cliffs, N. J. (1971), Sec. 2.35.

To prove the uniqueness of the "expansion" (1), suppose there were another expansion

$$\mathbf{x} = x'_1\mathbf{e}_1 + x'_2\mathbf{e}_2 + \cdots + x'_n\mathbf{e}_n$$

of $\mathbf{x}$ with respect to the vectors $\mathbf{e}_1, \mathbf{e}_2, \ldots, \mathbf{e}_n$, so that

$$x_1\mathbf{e}_1 + x_2\mathbf{e}_2 + \cdots + x_n\mathbf{e}_n = x'_1\mathbf{e}_1 + x'_2\mathbf{e}_2 + \cdots + x'_n\mathbf{e}_n,$$

and hence

$$(x_1 - x'_1)\mathbf{e}_1 + (x_2 - x'_2)\mathbf{e}_2 + \cdots + (x_n - x'_n)\mathbf{e}_n = \mathbf{0}.$$

But then

$$x_1 = x'_1, \quad x_2 = x'_2, \ldots, \quad x_n = x'_n,$$

because of the linear independence of $\mathbf{e}_1, \mathbf{e}_2, \ldots, \mathbf{e}_n$. ∎

Remark 1. A system of linearly independent vectors $\mathbf{e}_1, \mathbf{e}_2, \ldots, \mathbf{e}_n$ is called a *basis* for the n-dimensional space L if every vector $\mathbf{x} \in L$ has a unique expansion of the form (1), and the numbers $x_1, x_2, \ldots, x_n$ are then called the *components* of $\mathbf{x}$ with respect to (or relative to) this basis. Thus we have proved that *any n linearly independent vectors of L can be chosen as a basis for L.*

Remark 2. In particular, any vector $\mathbf{x}$ on the line L_1 has a unique representation of the form

$$\mathbf{x} = x_1\mathbf{e}_1,$$

where $\mathbf{e}_1$ is an arbitrary nonzero vector on the line, while any vector $\mathbf{x}$ in the plane has a unique representation of the form

$$\mathbf{x} = x_1\mathbf{e}_1 + x_2\mathbf{e}_2,$$

where $\mathbf{e}_1$ and $\mathbf{e}_2$ are any two noncollinear vectors of the plane. Similarly, any vector $\mathbf{x}$ in the space L_3 has a unique representation of the form

$$\mathbf{x} = x_1\mathbf{e}_1 + x_2\mathbf{e}_2 + x_3\mathbf{e}_3,$$

where $\mathbf{e}_1, \mathbf{e}_2$ and $\mathbf{e}_3$ are any three noncoplanar spatial vectors. Thus any vector in the space L_1, L_2 or L_3 is completely determined by its components with respect to an appropriate basis. Moreover, vectors in L_1, L_2 and L_3 have the following familiar properties:

a) Two vectors are equal if and only if their corresponding components are equal;
b) Each component of the sum of two vectors equals the sum of the corresponding components of the separate vectors;
c) Each component of the product of a number and a vector equals the product of the number and the corresponding component.

Hence it is clear that the spaces L_1, L_2 and L_3 can be regarded as the special cases of the space L_n obtained for $n = 1$, 2 and 3, respectively.

Remark 3. The expansion (1) can be written more concisely in the form

$$\mathbf{x} = \sum_{k=1}^{n} x_k \mathbf{e}_k. \tag{2}$$

But even this notation is not very convenient and can be simplified still further by dropping the summation sign, i.e., by writing

$$\mathbf{x} = x_k \mathbf{e}_k$$

instead of (2), *it being understood that summation from* 1 *to n is carried out over any index (in this case k) which appears twice in the same expression.* This rule, proposed by Einstein, is called the *summation convention*, and the index k is called an *index of summation*. We can just as well replace k by any other letter, so that

$$x_k \mathbf{e}_k = x_i \mathbf{e}_i = x_\alpha \mathbf{e}_\alpha = \cdots.$$

In the rest of the book, we will confine ourselves (for simplicity) to the case of the plane or ordinary three-dimensional space. Hence $n = 2$ or $n = 3$ in all subsequent formulas, so that indices of summation will range over the values 1 and 2 or over the values 1, 2 and 3. However, most of the considerations given below will remain valid for a general n-dimensional linear space.

PROBLEMS

1. Prove that the vectors

$$\mathbf{a}_1 = (1, 1, 1), \qquad \mathbf{a}_2 = (1, 1, 2), \qquad \mathbf{a}_3 = (1, 2, 3)$$

form a basis for the space L_3. Write the vector $\mathbf{x} = (6, 9, 14)$ in this basis.

2. Find the dimension of the space of all polynomials of degree not exceeding n (see Sec. 1, Example 6 and Sec. 2, Prob. 5). Find a basis for the space. What are the components of an arbitrary polynomial of the space with respect to this basis?

3. What is the dimension of the space $C[a, b]$?†

4. What is the dimension of the space R^+ considered in Sec. 1, Prob. 4? Find a basis in R^+.

5. Prove that if the dimension of a subspace L' of a finite-dimensional linear space L coincides with that of L itself, then $L' \equiv L$.

6. Prove that the sum of the dimensions of two linear subspaces L' and L'' of a finite-dimensional linear space L equals the dimension of $L' + L''$ (the sum of L' and L'') plus the dimension of $L' \cap L''$ (the intersection of L' and L'').

† See Sec. 1, Example 7 and Sec. 2, Prob. 6.

7. Prove that if the dimension of the sum $L' + L''$ of two linear subspaces L' and L'' of a finite-dimensional linear space L is one greater than the dimension of the intersection $L' \cap L''$, then $L' + L''$ coincides with one of the subspaces and $L' \cap L''$ with the other.

8. Prove that if two linear subspaces L' and L'' of a finite-dimensional linear space L have only the zero vector in common, then the dimension of $L' + L''$ cannot exceed that of L.

9. Describe the sum and intersection of two (distinct) two-dimensional linear subspaces of the space L_3.

10. By the *linear subspace* $L' \subset L_n$ *spanned by the vectors* $\mathbf{a}_1, \mathbf{a}_2, \ldots, \mathbf{a}_m$ is meant the linear subspace of smallest dimension containing $\mathbf{a}_1, \mathbf{a}_2, \ldots, \mathbf{a}_m$.† Let L' be the linear subspace of L_4 spanned by the vectors

$$\mathbf{a}_1 = (1, 1, 1, 1), \qquad \mathbf{a}_2 = (1, -1, 1, -1), \qquad \mathbf{a}_3 = (1, 3, 1, 3),$$

and let L'' be the linear subspace of L_4 spanned by the vectors

$$\mathbf{b}_1 = (1, 2, 0, 2), \qquad \mathbf{b}_2 = (1, 2, 1, 2), \qquad \mathbf{b}_3 = (3, 1, 3, 1).$$

Find the dimension s of the sum $L' + L''$ and the dimension d of the intersection $L' \cap L''$.

11. Let L' and L'' be the linear subspaces of L_4 spanned by the vectors

$$\mathbf{a}_1 = (1, 2, 1, -2), \qquad \mathbf{a}_2 = (2, 3, 1, 0), \qquad \mathbf{a}_3 = (1, 2, 2, -3)$$

and

$$\mathbf{b}_1 = (1, 1, 1, 1), \qquad \mathbf{b}_2 = (1, 0, 1, -1), \qquad \mathbf{b}_3 = (1, 3, 0, -4),$$

respectively. Find bases for $L' + L''$ and $L' \cap L''$.

12. Find a basis for each of the following subspaces of the space L_n:

a) The set of n-dimensional vectors whose first and second components are equal;
b) The set of n-dimensional vectors whose even-numbered components are equal;
c) The set of n-dimensional vectors of the form $(\alpha, \beta, \alpha, \beta, \ldots)$;
d) The set of n-dimensional vectors $(x_1, x_2, \ldots, x_n)$ such that $x_1 + x_2 + \cdots + x_n = 0$.

What is the dimension of each subspace?

13. Which solutions of a homogeneous linear differential equation of order n form a basis for the linear space of solutions of the equation (see Sec. 1, Prob. 5)? What is the dimension of this space? What numbers serve as components of an arbitrary solution of the equation with respect to the basis?

14. Write a single vector equation equivalent to the system of equations

$$\begin{aligned} a_{11}x_1 + a_{12}x_2 + \cdots + a_{1n}x_n &= b_1, \\ a_{21}x_1 + a_{22}x_2 + \cdots + a_{2n}x_n &= b_2, \\ \cdots\cdots\cdots\cdots\cdots\cdots\cdots & \\ a_{m1}x_1 + a_{m2}x_2 + \cdots + a_{mn}x_n &= b_m. \end{aligned}$$

† As usual, the symbol $\subset$ means "is a subset of."

4. Orthonormal Bases. The Scalar Product

In the three-dimensional space L_3, let $\mathbf{e}_1, \mathbf{e}_2, \mathbf{e}_3$ be a basis consisting of three (pairwise) orthogonal unit vectors.† Such a basis is said to be *orthonormal.* Expanding an arbitrary vector $\mathbf{x}$ with respect to an orthonormal basis $\mathbf{e}_1, \mathbf{e}_2, \mathbf{e}_3$, we get

$$\mathbf{x} = x_i\mathbf{e}_i$$

(the summation convention is in force), where the numbers x_i are called the *rectangular* components of the vector $\mathbf{x}$.

An orthonormal basis $\mathbf{e}_1, \mathbf{e}_2, \mathbf{e}_3$ is called *right-handed* if the rotation through 90° carrying the vector $\mathbf{e}_1$ into the vector $\mathbf{e}_2$ appears to be counterclockwise when seen from the end of the vector $\mathbf{e}_3$. If the same rotation appears to be clockwise, the basis is called *left-handed.*

By the *scalar product* of two vectors $\mathbf{x}$ and $\mathbf{y}$, denoted by $\mathbf{x}\cdot\mathbf{y}$ or $(\mathbf{x}, \mathbf{y})$, we mean the quantity

$$|\mathbf{x}||\mathbf{y}|\cos\theta,$$

where $|\mathbf{x}|$ is the length of the vector $\mathbf{x}$, $|\mathbf{y}|$ is the length of $\mathbf{y}$, and θ is the angle between $\mathbf{x}$ and $\mathbf{y}$ (varying between 0 and 180°). It is easy to see that the scalar product has the following properties:

1) $\mathbf{x}\cdot\mathbf{y} = \mathbf{y}\cdot\mathbf{x}$;
2) $(\lambda\mathbf{x})\cdot\mathbf{y} = \lambda\mathbf{x}\cdot\mathbf{y}$ for arbitrary real λ;
3) $(\mathbf{x} + \mathbf{y})\cdot\mathbf{z} = \mathbf{x}\cdot\mathbf{z} + \mathbf{y}\cdot\mathbf{z}$;‡
4) $\mathbf{x}\cdot\mathbf{x} \geq 0$ where $\mathbf{x}\cdot\mathbf{x} = 0$ if and only if $\mathbf{x} = \mathbf{0}$.

Let $\mathbf{e}_1, \mathbf{e}_2, \mathbf{e}_3$ be an orthonormal basis. Then the various scalar products of the vectors $\mathbf{e}_1, \mathbf{e}_2, \mathbf{e}_3$ with each other are given by the following table:

	$\mathbf{e}_1$	$\mathbf{e}_2$	$\mathbf{e}_3$
$\mathbf{e}_1$	1	0	0
$\mathbf{e}_2$	0	1	0
$\mathbf{e}_3$	0	0	1

Introducing the quantity δ_{ij} defined by

$$\delta_{ij} = \begin{cases} 1 \text{ if } i = j, \\ 0 \text{ if } i \neq j, \end{cases}$$

we find that

$$\mathbf{e}_i\cdot\mathbf{e}_j = \delta_{ij} \qquad (i, j = 1, 2, 3).$$

We call δ_{ij} the *symmetric Kronecker symbol*, or simply the *Kronecker delta.*

† By a *unit vector* is meant a vector of unit length.

‡ The simplest way of proving property 3) is to use formula (1) below.

Next let $\mathbf{x} = x_i\mathbf{e}_i$ and $\mathbf{y} = y_j\mathbf{e}_j$ be two arbitrary vectors of L_3. Then

$$\mathbf{x}\cdot\mathbf{y} = (x_i\mathbf{e}_i)\cdot(y_j\mathbf{e}_j),$$

and hence

$$\mathbf{x}\cdot\mathbf{y} = x_iy_j(\mathbf{e}_i\cdot\mathbf{e}_j)$$

(why?). The sum on the right consists of nine terms, since the indices i and j range independently from 1 to 3. But only three of these terms are nonzero, since $\mathbf{e}_i\cdot\mathbf{e}_j = 0$ if $i \neq j$. Moreover $\mathbf{e}_i\cdot\mathbf{e}_j = 1$ if $i = j$, and hence

$$\mathbf{x}\cdot\mathbf{y} = x_1y_1 + x_2y_2 + x_3y_3, \tag{1}$$

which can be written more concisely in the form

$$\mathbf{x}\cdot\mathbf{y} = x_iy_i$$

by using the summation convention.

Remark. The scalar product of an arbitrary vector $\mathbf{x} = x_i\mathbf{e}_i$ and the basis vector $\mathbf{e}_j$ is clearly

$$\mathbf{x}\cdot\mathbf{e}_j = x_i(\mathbf{e}_i\cdot\mathbf{e}_j) = x_i\delta_{ij},$$

where the expression $x_i\delta_{ij}$ is the sum of three terms, two of which vanish since $\delta_{ij} = 0$ if $i \neq j$. But $\delta_{ij} = 1$ if $i = j$, and hence

$$\mathbf{x}\cdot\mathbf{e}_j = x_i\delta_{ij} = x_j.$$

Thus the rectangular components of the vector $\mathbf{x}$ are the orthogonal projections of $\mathbf{x}$ onto the corresponding coordinate axes.

Finally, we list a number of familiar geometric facts involving scalar products:

a) The *length* of the vector $\mathbf{x} = x_i\mathbf{e}_i$ is given by

$$|\mathbf{x}| = \sqrt{\mathbf{x}\cdot\mathbf{x}} = \sqrt{\delta_{ij}x_ix_j} = \sqrt{x_ix_i}.$$

Any vector $\mathbf{x} \neq \mathbf{0}$ can be *normalized*, i.e., replaced by a proportional vector $\mathbf{x}_0$ of unit length, by merely setting

$$\mathbf{x}_0 = \frac{\mathbf{x}}{|\mathbf{x}|}.$$

b) The cosine of the angle θ between the vectors $\mathbf{x} = x_i\mathbf{e}_i$ and $\mathbf{y} = y_i\mathbf{e}_i$ is given by

$$\cos\theta = \frac{\mathbf{x}\cdot\mathbf{y}}{\sqrt{|\mathbf{x}|\,|\mathbf{y}|}} = \frac{x_iy_i}{\sqrt{x_ix_i}\sqrt{y_iy_i}}.$$

c) If $\mathbf{a}$ is a unit vector, then its ith component a_i equals the cosine of the angle α_i which $\mathbf{a}$ makes with the basis vector $\mathbf{e}_i$, i.e.,

$$a_i = \mathbf{a}\cdot\mathbf{e}_i = \cos\alpha_i.$$

Moreover

$$\cos^2\alpha_1 + \cos^2\alpha_2 + \cos^2\alpha_3 = 1,$$

since $\mathbf{a}\cdot\mathbf{a} = 1$.

d) The *projection* of the vector $\mathbf{x} = x_i \mathbf{e}_i$ onto the vector $\mathbf{a} = a_i \mathbf{e}_i$ is given by

$$\mathrm{Pr}_{\mathbf{a}}\mathbf{x} = \frac{\mathbf{a}\cdot\mathbf{x}}{|\mathbf{a}|} = \frac{a_i x_i}{\sqrt{a_i a_i}}.$$

PROBLEMS

1. Use the scalar product to prove the following theorems of elementary geometry:

a) The cosine law for a triangle;
b) The sum of the squares of the diagonals of a parallelogram equals twice the sum of the squares of two adjacent sides of the parallelogram;
c) The diagonals of a rhombus are perpendicular;
d) The diagonals of a rectangle are equal;
e) The Pythagorean theorem;
f) The length m_a of the median of a triangle with sides a, b, and c equals

$$m_a = \sqrt{\frac{b^2 + c^2}{2} - \frac{a^2}{4}};$$

g) If two medians of a triangle are equal, then the triangle is isosceles;
h) The sum of the squares of the diagonals of a trapezoid equals the sum of the squares of the lateral sides plus twice the product of the bases;
i) Opposite edges of a regular tetrahedron are orthogonal.

2. Prove the *Cauchy–Schwarz inequality*

$$(\mathbf{x}\cdot\mathbf{y})^2 \le |\mathbf{x}|^2 |\mathbf{y}|^2, \tag{2}$$

and write it in terms of the components of the vectors $\mathbf{x}, \mathbf{y} \in L_3$. Prove that the equality holds only if $\mathbf{x}$ and $\mathbf{y}$ are collinear.

3. Prove the following *triangle inequalities* involving vectors $\mathbf{x}, \mathbf{y} \in L_3$:

$$|\mathbf{x} + \mathbf{y}| \le |\mathbf{x}| + |\mathbf{y}|, \qquad |\mathbf{x} - \mathbf{y}| \ge ||\mathbf{x}| - |\mathbf{y}||. \tag{3}$$

4. Given an arbitrary linear space L, we say that a scalar product is defined in L if with every pair of vectors $\mathbf{x}, \mathbf{y} \in L$ there is associated a number $\mathbf{x}\cdot\mathbf{y}$, or equivalently $(\mathbf{x}, \mathbf{y})$, such that

a) $\mathbf{x}\cdot\mathbf{y} = \mathbf{y}\cdot\mathbf{x}$;
b) $(\lambda\mathbf{x})\cdot\mathbf{y} = \lambda(\mathbf{x}\cdot\mathbf{y})$ for arbitrary real λ;
c) $(\mathbf{x} + \mathbf{y})\cdot\mathbf{z} = \mathbf{x}\cdot\mathbf{z} + \mathbf{y}\cdot\mathbf{z}$;
d) $\mathbf{x}\cdot\mathbf{x} \ge 0$ where $\mathbf{x}\cdot\mathbf{x} = 0$ if and only if $\mathbf{x} = \mathbf{0}$.

A linear space equipped with a scalar product is called a *Euclidean space*, conventionally denoted by E. The concepts of the length of a vector and of the angle between two vectors in a Euclidean space E are defined by analogy with the corresponding concepts in the three-dimensional Euclidean space L_3 (or E_3) considered above. Thus the *length* of a vector $\mathbf{x} \in E$ is defined as

$$|\mathbf{x}| = \sqrt{\mathbf{x}\cdot\mathbf{x}},$$

while the angle between two vectors $\mathbf{x}, \mathbf{y} \in E$ is the angle φ ($0° \leq \varphi \leq 180°$) whose cosine equals

$$\cos \varphi = \frac{\mathbf{x} \cdot \mathbf{y}}{|\mathbf{x}||\mathbf{y}|}.$$

Two vectors $\mathbf{x}, \mathbf{y} \in E$ are said to be *orthogonal* if $\mathbf{x} \cdot \mathbf{y} = 0$.

Can the scalar product of two vectors in L_3 be defined as

a) The product of their lengths;

b) The product of their lengths and the square of the cosine of the angle between them;

c) Three times the ordinary scalar product?

5. Prove that the scalar product of two vectors $\mathbf{x} = (x_1, \ldots, x_n)$ and $\mathbf{y} = (y_1, \ldots, y_n)$ in the space L_n (see Sec. 1, Example 5) can be defined by the formula

$$\mathbf{x} \cdot \mathbf{y} = x_1 y_1 + \cdots + x_n y_n. \tag{4}$$

(The space L_n equipped with this scalar product is a Euclidean space, which we denote by E_n.)

6. Prove that the scalar product of two functions $f(t)$ and $g(t)$ in the space $C[a, b]$ (see Sec. 1, Example 7) can be defined by the formula

$$(f, g) = \int_a^b f(t)g(t)\, dt.$$

Write an expression for the length of $f(t)$.

7. Prove that the scalar product of two arbitrary vectors $\mathbf{x} = (x_1, \ldots, x_n)$ and $\mathbf{y} = (y_1, \ldots, y_n)$ in the n-dimensional Euclidean space E_n is given by the expression (4) if and only if the underlying basis $\mathbf{e}_1, \ldots, \mathbf{e}_n$ (in which $x_i = \mathbf{x} \cdot \mathbf{e}_i$, $y_i = \mathbf{y} \cdot \mathbf{e}_i$) is orthonormal.

8. Prove that the Cauchy–Schwarz inequality (2) holds in an arbitrary Euclidean space.

9. Write the Cauchy–Schwarz inequality for vectors of the space E_n in component form and for vectors of the space $C[a, b]$ equipped with the scalar product defined in Prob. 6.

10. Find the angles of the triangle in the space $C[-1, 1]$ formed by the vectors $x_1(t) = 1$, $x_2(t) = t$, $x_3(t) = 1 - t$.

11. Prove that any two vectors of the system of trigonometric functions

$$1, \cos t, \sin t, \cos 2t, \sin 2t, \ldots, \cos nt, \sin nt, \ldots$$

in the space $C[-\pi, \pi]$ are orthogonal.

12. Prove that any n (pairwise) orthogonal nonzero vectors $\mathbf{x}_1, \ldots, \mathbf{x}_n \in E_n$ are linearly independent.

13. Prove that if the vector $\mathbf{x}$ in a Euclidean space E is orthogonal to the vectors $\mathbf{y}_1, \ldots, \mathbf{y}_k$, then $\mathbf{x}$ is orthogonal to any linear combination $c_1\mathbf{y}_1 + \cdots + c_k\mathbf{y}_k$.

14. Let $\mathbf{x}_1, \ldots, \mathbf{x}_n$ be the same as in Prob. 12. Prove that

$$|\mathbf{x}_1 + \mathbf{x}_2 + \cdots + \mathbf{x}_k|^2 = |\mathbf{x}_1|^2 + |\mathbf{x}_2|^2 + \cdots + |\mathbf{x}_k|^2,$$

thereby generalizing the Pythagorean theorem (cf. Prob. 1e).

15. Show that the triangle inequalities (3) hold for arbitrary vectors $\mathbf{x}$ and $\mathbf{y}$ in any Euclidean space E.

16. Write the triangle inequalities for the space $C[a, b]$ equipped with the scalar product defined in Prob. 6.

17. Let $\mathbf{e}_1, \ldots, \mathbf{e}_n$ be an orthonormal basis in E_n. Prove *Bessel's inequality*

$$\sum_{i=1}^{k} (\mathrm{Pr}_{\mathbf{e}_i}\mathbf{x})^2 \leq \mathbf{x}\cdot\mathbf{x} \qquad (k \leq n). \tag{5}$$

Prove that the equality holds if and only if $k = n$.†

18. Let E_{n+1} be the Euclidean space consisting of all polynomials of degree not exceeding n, with real coefficients, where the scalar product of the polynomials $P(t)$ and $Q(t)$ is defined by the formula

$$(P, Q) = \int_{-1}^{1} P(t)Q(t)\, dt.$$

a) Prove that the polynomials

$$P_0(t) = 1, \qquad P_k(t) = \frac{1}{2^k k!}\frac{d^k}{dt^k}(t^2 - 1)^k \qquad (k = 1, 2, \ldots, n),$$

known as *Legendre polynomials*, form an orthogonal basis in E_{n+1}.

b) Write the Legendre polynomials for $k = 0, 1, 2, 3, 4$. Verify that the degree of $P_k(t)$ is k, and expand $P_k(t)$ in powers of t.

c) What is the "length" of $P_k(t)$?

d) Find $P_k(1)$.

5. The Vector Product. Triple Products

5.1. By the *vector* (or *cross*) *product* of two vectors $\mathbf{x}$ and $\mathbf{y}$, denoted by $\mathbf{x} \times \mathbf{y}$, we mean the vector $\mathbf{z}$ such that

1) The length of $\mathbf{z}$ equals the area of the parallelogram constructed on the vectors $\mathbf{x}$ and $\mathbf{y}$, i.e., $|\mathbf{z}| = |\mathbf{x}||\mathbf{y}| \sin \varphi$ where φ is the angle between the vectors $\mathbf{x}$ and $\mathbf{y}$;
2) The vector $\mathbf{z}$ is orthogonal to each of the vectors $\mathbf{x}$ and $\mathbf{y}$;
3) The vectors $\mathbf{x}$, $\mathbf{y}$, and $\mathbf{z}$ (in that order) form a right-handed triple.

The following properties of the vector product are easily verified:

1) $\mathbf{x} \times \mathbf{y} = -(\mathbf{y} \times \mathbf{x})$;
2) $(\lambda\mathbf{x}) \times \mathbf{y} = \lambda(\mathbf{x} \times \mathbf{y})$;
3) $(\mathbf{x} + \mathbf{y}) \times \mathbf{z} = \mathbf{x} \times \mathbf{z} + \mathbf{y} \times \mathbf{z}$.

Let $\mathbf{e}_1, \mathbf{e}_2, \mathbf{e}_3$ be an orthonormal basis in the space L_3. Then the various vector products of the basis vectors with each other are given by the table

† In this case (5) reduces to *Parseval's theorem*

$$\sum_{i=1}^{n} (\mathrm{Pr}_{\mathbf{e}_i}\mathbf{x})^2 = \mathbf{x}\cdot\mathbf{x}.$$

	$\mathbf{e}_1$	$\mathbf{e}_2$	$\mathbf{e}_3$
$\mathbf{e}_1$	0	$\mathbf{e}_3$	$-\mathbf{e}_2$
$\mathbf{e}_2$	$-\mathbf{e}_3$	0	$\mathbf{e}_1$
$\mathbf{e}_3$	$\mathbf{e}_2$	$-\mathbf{e}_1$	0

if $\mathbf{e}_1, \mathbf{e}_2, \mathbf{e}_3$ is a right-handed basis, and by the table

	$\mathbf{e}_1$	$\mathbf{e}_2$	$\mathbf{e}_3$
$\mathbf{e}_1$	0	$-\mathbf{e}_3$	$\mathbf{e}_2$
$\mathbf{e}_2$	$\mathbf{e}_3$	0	$-\mathbf{e}_1$
$\mathbf{e}_3$	$-\mathbf{e}_2$	$\mathbf{e}_1$	0

if $\mathbf{e}_1, \mathbf{e}_2, \mathbf{e}_3$ is a left-handed basis.

To write the vector products of the basis vectors in a form valid for any orthonormal basis, we introduce a quantity ϵ, equal to $+1$ if the basis $\mathbf{e}_1, \mathbf{e}_2, \mathbf{e}_3$ is right-handed and to -1 if the basis is left-handed. Thus ϵ depends on the "handedness" of the basis. We then introduce a quantity ϵ_{ijk} given by

$$\epsilon_{123} = \epsilon_{231} = \epsilon_{312} = \epsilon,$$

$$\epsilon_{213} = \epsilon_{321} = \epsilon_{123} = -\epsilon$$

if all three subscripts are different, and equal to zero if any two of the indices i, j, k are equal. This quantity, which depends on the choice of the basis, is called the *antisymmetric Kronecker symbol.* Using ϵ_{ijk}, we have

$$\mathbf{e}_i \times \mathbf{e}_j = \epsilon_{ijk}\mathbf{e}_k, \tag{1}$$

regardless of the handedness of the basis, where in the right-hand side it is understood that summation is carried out over the index k (in keeping with the convention introduced on p. 10). Formula (1) is easily verified. For example, we have

$$\mathbf{e}_1 \times \mathbf{e}_2 = \epsilon_{12k}\mathbf{e}_k = \epsilon_{121}\mathbf{e}_1 + \epsilon_{122}\mathbf{e}_2 + \epsilon_{123}\mathbf{e}_3,$$

where the first two terms on the right vanish while $\epsilon_{123} = \epsilon$. It follows that

$$\mathbf{e}_1 \times \mathbf{e}_2 = \epsilon\mathbf{e}_3,$$

so that

$$\mathbf{e}_1 \times \mathbf{e}_2 = \mathbf{e}_3$$

if the basis is right-handed, while

$$\mathbf{e}_1 \times \mathbf{e}_2 = -\mathbf{e}_3$$

if the basis is left-handed, in keeping with the tables given above. More generally, given any two vectors $\mathbf{x} = x_i\mathbf{e}_i$ and $\mathbf{y} = y_j\mathbf{e}_j$, we have

$$\mathbf{x} \times \mathbf{y} = (x_i\mathbf{e}_i) \times (y_j\mathbf{e}_j),$$

and hence

$$\mathbf{x} \times \mathbf{y} = x_i y_j (\mathbf{e}_i \times \mathbf{e}_j) = \epsilon_{ijk} x_i y_j \mathbf{e}_k$$

(why?), where in the right-hand side summation takes place over all three indices. In detail,

$$\mathbf{x} \times \mathbf{y} = \epsilon\{(x_2 y_3 - x_3 y_2)\mathbf{e}_1 + (x_3 y_1 - x_1 y_3)\mathbf{e}_2 + (x_1 y_2 - x_2 y_1)\mathbf{e}_3\},$$

or more concisely,

$$\mathbf{x} \times \mathbf{y} = \epsilon \begin{vmatrix} \mathbf{e}_1 & \mathbf{e}_2 & \mathbf{e}_3 \\ x_1 & x_2 & x_3 \\ y_1 & y_2 & y_3 \end{vmatrix}$$

in terms of a third-order determinant. Thus, if $\mathbf{z} = \mathbf{x} \times \mathbf{y}$, the components of the vector $\mathbf{z}$ are given by

$$x_k = \epsilon_{kij} x_i y_j$$

(since $\epsilon_{ijk} = \epsilon_{kij}$), or in more detail,

$$z_1 = \epsilon(x_2 y_3 - x_3 y_2),$$
$$z_2 = \epsilon(x_3 y_1 - x_1 y_3),$$
$$z_3 = \epsilon(x_1 y_2 - x_2 y_1).$$

Remark. It should be noted that our definition of a vector product differs somewhat from another definition often encountered in the literature.† With our definition the vector product is independent of the handedness of the underlying basis, while in the alternative definition the vector product changes sign whenever the handedness of the basis is changed and hence is not an ordinary vector but rather a so-called "axial vector." However, as defined here, the vector product is an ordinary vector, a fact which frees us from the necessity of considering axial vectors.

5.2. The *scalar triple product* $(\mathbf{x}, \mathbf{y}, \mathbf{z})$ of three vectors $\mathbf{x}$, $\mathbf{y}$ and $\mathbf{z}$ is defined by the formula

$$(\mathbf{x}, \mathbf{y}, \mathbf{z}) = (\mathbf{x} \times \mathbf{y}) \cdot \mathbf{z},$$

and equals the volume of the parallelepiped constructed on the vectors $\mathbf{x}$, $\mathbf{y}$ and $\mathbf{z}$, taken with the plus sign if the vectors $\mathbf{x}$, $\mathbf{y}$ and $\mathbf{z}$ (in that order) form a right-handed triple and with the minus sign otherwise. The scalar triple product has the following easily verified properties:

1) $(\mathbf{x}, \mathbf{y}, \mathbf{z}) = -(\mathbf{y}, \mathbf{x}, \mathbf{z})$;
2) $(\mathbf{x}, \mathbf{y}, \mathbf{z}) = (\mathbf{y}, \mathbf{z}, \mathbf{x}) = (\mathbf{z}, \mathbf{x}, \mathbf{y})$;‡

† See e.g., A. I. Borisenko and I. E. Tarapov, *Vector and Tensor Analysis with Applications*, translated by R. A. Silverman, Prentice-Hall, Inc., Englewood Cliffs, N. J. (1968), p. 18.

‡ The simplest way of proving property 2) is to use the geometric interpretation of the scalar triple product. Alternatively use formula (5) below, together with a familiar property of determinants.

3) $(\lambda\mathbf{x}, \mathbf{y}, \mathbf{z}) = \lambda(\mathbf{x}, \mathbf{y}, \mathbf{z})$;
4) $(\mathbf{x} + \mathbf{y}, \mathbf{z}, \mathbf{u}) = (\mathbf{x}, \mathbf{z}, \mathbf{u}) + (\mathbf{y}, \mathbf{z}, \mathbf{u})$.

Moreover, if $\mathbf{e}_i, \mathbf{e}_j, \mathbf{e}_k$ are basis vectors, then

$$(\mathbf{e}_i, \mathbf{e}_j, \mathbf{e}_k) = \epsilon_{ijk}. \tag{2}$$

In fact,

$$(\mathbf{e}_i, \mathbf{e}_j, \mathbf{e}_k) = (\mathbf{e}_i \times \mathbf{e}_j)\cdot\mathbf{e}_k = \epsilon_{ijl}\mathbf{e}_l\cdot\mathbf{e}_k, \tag{3}$$

where the right-hand side involves summation over the index l. But $\mathbf{e}_l\cdot\mathbf{e}_k$ is nonzero only if $l = k$, in which case $\mathbf{e}_l\cdot\mathbf{e}_k = 1$. Hence the right-hand side of (3) reduces to the single term ϵ_{ijk}, thereby proving (2).

Now let $\mathbf{x} = x_i\mathbf{e}_i$, $\mathbf{y} = y_j\mathbf{e}_j$, $\mathbf{z} = z_k\mathbf{e}_k$ be three arbitrary vectors. Then

$$(\mathbf{x}, \mathbf{y}, \mathbf{z}) = (x_i\mathbf{e}_i, y_j\mathbf{e}_j, z_k\mathbf{e}_k) = x_iy_jz_k(\mathbf{e}_i, \mathbf{e}_j, \mathbf{e}_k)$$

(why?), and hence

$$(\mathbf{x}, \mathbf{y}, \mathbf{z}) = \epsilon_{ijk}x_iy_jz_k, \tag{4}$$

where the right-hand side involves summation over the indices i, j and k which independently range from 1 to 3. Thus the expression on the right is a sum containing $3^3 = 27$ terms, of which only six are nonzero since the other terms involve ϵ_{ijk} with repeated indices. Hence, writing (4) out in full, we get

$$(\mathbf{x}, \mathbf{y}, \mathbf{z}) = \epsilon(x_1y_2z_3 + x_2y_3z_1 + x_3y_1z_2 - x_2y_1z_3 - x_3y_2z_1 - x_1y_3z_2),$$

or, more concisely,

$$(\mathbf{x}, \mathbf{y}, \mathbf{z}) = \epsilon\begin{vmatrix} x_1 & x_2 & x_3 \\ y_1 & y_2 & y_3 \\ z_1 & z_2 & z_3 \end{vmatrix} \tag{5}$$

in terms of a third-order determinant.

5.3. Finally we consider the *vector triple product* $\mathbf{x} \times (\mathbf{y} \times \mathbf{z})$ of three vectors $\mathbf{x}$, $\mathbf{y}$ and $\mathbf{z}$, establishing the formula

$$\mathbf{x} \times (\mathbf{y} \times \mathbf{z}) = \mathbf{y}(\mathbf{x}\cdot\mathbf{z}) - \mathbf{z}(\mathbf{x}\cdot\mathbf{y}). \tag{6}$$

If the vectors $\mathbf{y}$ and $\mathbf{z}$ are collinear, then it is easy to see that both sides of (6) vanish. Thus suppose that $\mathbf{y}$ and $\mathbf{z}$ are not collinear, and let $\mathbf{u} = \mathbf{x} \times (\mathbf{y} \times \mathbf{z})$. The vector $\mathbf{u}$ is orthogonal to the vector $\mathbf{y} \times \mathbf{z}$, and hence lies in the plane Π determined by the vectors $\mathbf{y}$ and $\mathbf{z}$, i.e.,

$$\mathbf{u} = \lambda\mathbf{y} + \mu\mathbf{z}. \tag{7}$$

Let $\mathbf{z}^*$ denote the vector in the plane Π obtained by rotating $\mathbf{z}$ through 90° in the clockwise direction as seen from the end of the vector $\mathbf{y} \times \mathbf{z}$. Then the vectors $\mathbf{z}^*$, $\mathbf{z}$ and $\mathbf{y} \times \mathbf{z}$ form a right-handed triple of vectors, and clearly

$$\mathbf{u}\cdot\mathbf{z}^* = \alpha(\mathbf{y}\cdot\mathbf{z}^*). \tag{8}$$

On the other hand,

$$\mathbf{u}\cdot\mathbf{z}^* = [\mathbf{x} \times (\mathbf{y} \times \mathbf{z})]\cdot\mathbf{z}^* = [(\mathbf{y} \times \mathbf{z}) \times \mathbf{z}^*]\cdot\mathbf{x},$$

by property 2) of the scalar triple product. Let $\mathbf{v} = (\mathbf{y} \times \mathbf{z}) \times \mathbf{z}^*$. Then $\mathbf{v}$ has the same direction as the vector $\mathbf{z}$, and moreover $|\mathbf{v}| = |\mathbf{y} \times \mathbf{z}||\mathbf{z}^*|$ since the vectors $\mathbf{y} \times \mathbf{z}$ and $\mathbf{z}$ are orthogonal. It follows that

$$|\mathbf{v}| = |\mathbf{y}||\mathbf{z}|^2 \sin(\mathbf{y}, \mathbf{z}) = |\mathbf{y}||\mathbf{z}|^2 \cos(\mathbf{y}, \mathbf{z}^*) = (\mathbf{y}\cdot\mathbf{z}^*)|\mathbf{z}|,$$

where $(\mathbf{y}, \mathbf{z})$ denotes the angle between $\mathbf{y}$ and $\mathbf{z}$. Therefore

$$\mathbf{v} = (\mathbf{y}\cdot\mathbf{z}^*)\mathbf{z},$$

and hence

$$\mathbf{u}\cdot\mathbf{z}^* = (\mathbf{x}\cdot\mathbf{z})(\mathbf{y}\cdot\mathbf{z}^*). \tag{9}$$

Comparing (8) and (9), we find that $\lambda = \mathbf{x}\cdot\mathbf{z}$. Moreover, taking the scalar product of (7) with the vector $\mathbf{x}$, we get

$$\lambda(\mathbf{x}\cdot\mathbf{y}) + \mu(\mathbf{x}\cdot\mathbf{z}) = 0,$$

which implies $\mu = -\mathbf{x}\cdot\mathbf{y}$. Substituting these values of λ and μ into (7), we finally arrive at (6).

PROBLEMS

1. Find the areas of the diagonal sections of the parallelepiped constructed on the vectors $\mathbf{a}$, $\mathbf{b}$ and $\mathbf{c}$.

2. Express the sine of the dihedral angle α formed at the edge AB of the tetrahedron $OABC$ in terms of the vectors $\overrightarrow{OA}$, $\overrightarrow{OB}$ and $\overrightarrow{OC}$.

3. Express the altitudes h_1, h_2, h_3 of a triangle in terms of the radius vectors $\mathbf{r}_1, \mathbf{r}_2, \mathbf{r}_3$ of its vertices.

4. Prove that the sum of the normal vectors $\mathbf{n}_1, \ldots, \mathbf{n}_4$ to the faces of a tetrahedron $OABC$, directed outwards from the tetrahedron and equal to the areas of the corresponding faces, equals zero. Prove that the areas $S_1, \ldots, S_4$ of the faces satisfy the formula

$$S_4^2 = S_1^2 + S_2^2 + S_3^2 - 2S_1S_2\cos(S_1, S_2) - 2S_2S_3\cos(S_2, S_3) - 2S_3S_1\cos(S_3, S_1),$$

where (S_i, S_j) denotes the angle between the faces with areas S_i and S_j.

5. Given a determinant

$$a = \begin{vmatrix} a_{11} & a_{12} & a_{13} \\ a_{21} & a_{22} & a_{23} \\ a_{31} & a_{32} & a_{33} \end{vmatrix}$$

of order three, let A_{ij} be the cofactor of the element a_{ij}. Prove that

a) $a = \dfrac{1}{3!}\epsilon_{ijk}\epsilon_{pqr}a_{ip}a_{jq}a_{kr}$;

b) $A_{ij} = \dfrac{1}{2!}\epsilon_{ikl}\epsilon_{jpq}a_{kp}a_{lq}$;

c) $A_{ik}a_{kj} = \delta_{ij}a$.

6. Prove *Lagrange's identity*

$$(\mathbf{a} \times \mathbf{b})\cdot(\mathbf{c} \times \mathbf{d}) = \begin{vmatrix} \mathbf{a}\cdot\mathbf{c} & \mathbf{a}\cdot\mathbf{d} \\ \mathbf{b}\cdot\mathbf{c} & \mathbf{b}\cdot\mathbf{d} \end{vmatrix}.$$

7. Use the result of the preceding problem to find $|\mathbf{a} \times \mathbf{b}|^2$, writing the result in component form.

8. Prove that the vectors **a** and **c** are collinear if

$$\mathbf{a} \times (\mathbf{b} \times \mathbf{c}) = (\mathbf{a} \times \mathbf{b}) \times \mathbf{c}, \qquad \mathbf{a}\cdot\mathbf{b} \neq 0, \qquad \mathbf{b}\cdot\mathbf{c} \neq 0.$$

9. Prove *Jacobi's identity*

$$\mathbf{a} \times (\mathbf{b} \times \mathbf{c}) + \mathbf{b} \times (\mathbf{c} \times \mathbf{a}) + \mathbf{c} \times (\mathbf{a} \times \mathbf{b}) = \mathbf{0}.$$

10. Suppose that in each face of a trihedral angle we draw a line through the vertex of the angle perpendicular to the edge lying opposite the face. Prove that the resulting three lines are coplanar. (It is assumed that none of the edges of the trihedral angle is perpendicular to the opposite face.)

11. Given four vectors **a**, **b**, **c** and **d** emanating from a common point, suppose **a** and **b** are orthogonal, while **c** and **d** are also orthogonal. Prove that the vectors $\mathbf{p} = (\mathbf{b} \times \mathbf{c}) \times (\mathbf{a} \times \mathbf{d})$ and $\mathbf{q} = (\mathbf{a} \times \mathbf{c}) \times (\mathbf{b} \times \mathbf{d})$ are orthogonal.

12. Find the area S of the base of a triangular pyramid, given the lengths a, b, c of the lateral edges and planar angles α, β, γ at the vertex (α lies opposite a, etc.).

13. Calculate the scalar triple product $(\mathbf{a} + \mathbf{b}, \mathbf{b} + \mathbf{c}, \mathbf{c} + \mathbf{a})$ and interpret the result geometrically.

14. Given three noncoplanar vectors **a**, **b** and **c**, what relation between the numbers λ, μ and ν makes the vectors $\mathbf{a} + \lambda\mathbf{b}$, $\mathbf{b} + \mu\mathbf{c}$ and $\mathbf{c} + \nu\mathbf{a}$ coplanar? Use the result to prove the *direct theorem of Menelaus* (the product of the ratios in which any line divides the sides of a triangle equals -1) and the *inverse theorem of Menelaus* (if three points lying on the sides of a triangle divide them in ratios whose product equals -1, then the three points lie on a line).

15. Use the scalar triple product to deduce *Cramer's theorem* for solving a system of three linear equations in three unknowns, written in vector form (cf. Sec. 3, Prob. 14).

16. Use formula (6) to prove the following formulas:

a) $(\mathbf{a} \times \mathbf{b}) \times (\mathbf{c} \times \mathbf{d}) = \mathbf{b}(\mathbf{a}, \mathbf{c}, \mathbf{d}) - \mathbf{a}(\mathbf{b}, \mathbf{c}, \mathbf{d})$;

b) $(\mathbf{a} \times \mathbf{b}, \mathbf{c} \times \mathbf{d}, \mathbf{e} \times \mathbf{f}) = (\mathbf{b}, \mathbf{e}, \mathbf{f})(\mathbf{a}, \mathbf{c}, \mathbf{d}) - (\mathbf{a}, \mathbf{e}, \mathbf{f})(\mathbf{b}, \mathbf{c}, \mathbf{d})$.

17. Prove that

a) $\mathbf{a}(\mathbf{b}, \mathbf{c}, \mathbf{d}) - \mathbf{b}(\mathbf{c}, \mathbf{d}, \mathbf{a}) + \mathbf{c}(\mathbf{d}, \mathbf{a}, \mathbf{b}) - \mathbf{d}(\mathbf{a}, \mathbf{b}, \mathbf{c}) = \mathbf{0}$;

b) $(\mathbf{a} \times \mathbf{b}, \mathbf{b} \times \mathbf{c}, \mathbf{c} \times \mathbf{a}) = |(\mathbf{a}, \mathbf{b}, \mathbf{c})|^2$.

What is the geometric meaning of the second formula?

18. Prove that

$$(\mathbf{a}, \mathbf{b}, \mathbf{c})(\mathbf{x}, \mathbf{y}, \mathbf{z}) = \begin{vmatrix} \mathbf{a}\cdot\mathbf{x} & \mathbf{a}\cdot\mathbf{y} & \mathbf{a}\cdot\mathbf{z} \\ \mathbf{b}\cdot\mathbf{x} & \mathbf{b}\cdot\mathbf{y} & \mathbf{b}\cdot\mathbf{z} \\ \mathbf{c}\cdot\mathbf{x} & \mathbf{c}\cdot\mathbf{y} & \mathbf{c}\cdot\mathbf{z} \end{vmatrix}.$$

19. Prove that three vectors making angles of α, β, γ with each other are coplanar if and only if

$$\begin{vmatrix} 1 & \cos\beta & \cos\gamma \\ \cos\beta & 1 & \cos\alpha \\ \cos\gamma & \cos\alpha & 1 \end{vmatrix} = 0.$$

6. Basis Transformations. Tensor Calculus

6.1. Let $\mathbf{e}_1, \mathbf{e}_2, \mathbf{e}_3$ be an orthonormal basis in the space L_3 and let $\mathbf{e}_{1'}, \mathbf{e}_{2'}, \mathbf{e}_{3'}$ be another orthonormal basis in L_3, both emanating from the same origin O (see Figure 3). Clearly, the vectors of the "new" basis $\mathbf{e}_{1'}, \mathbf{e}_{2'}, \mathbf{e}_{3'}$ can be expressed as linear combinations of the vectors of the "old" basis $\mathbf{e}_1, \mathbf{e}_2, \mathbf{e}_3$. Let $\gamma_{i'i}$ denote the coefficient of $\mathbf{e}_i$ in the expansion of $\mathbf{e}_{i'}$ with respect to the old basis vectors. Then the expansions of the new basis vectors with respect to the old basis vectors take the form

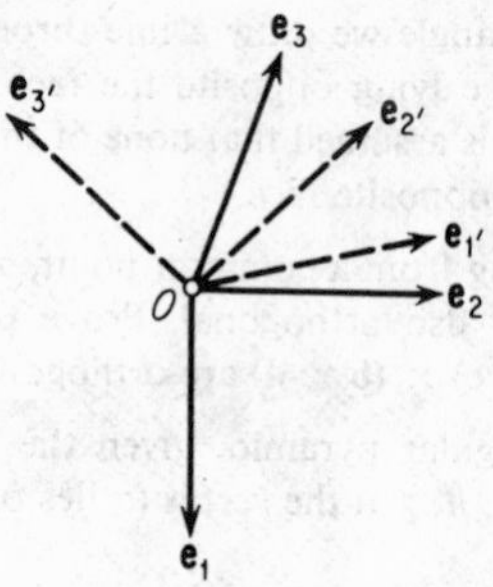

FIGURE 3

$$\mathbf{e}_{1'} = \gamma_{1'1}\mathbf{e}_1 + \gamma_{1'2}\mathbf{e}_2 + \gamma_{1'3}\mathbf{e}_3,$$
$$\mathbf{e}_{2'} = \gamma_{2'1}\mathbf{e}_1 + \gamma_{2'2}\mathbf{e}_2 + \gamma_{2'3}\mathbf{e}_3,$$
$$\mathbf{e}_{3'} = \gamma_{3'1}\mathbf{e}_1 + \gamma_{3'2}\mathbf{e}_2 + \gamma_{3'3}\mathbf{e}_3,$$

or more concisely,

$$\mathbf{e}_{i'} = \gamma_{i'i}\mathbf{e}_i. \tag{1}$$

Taking the scalar product of each of the equations (1) with each of the vectors $\mathbf{e}_j$, we get

$$\mathbf{e}_{i'}\cdot\mathbf{e}_j = \gamma_{i'i}\mathbf{e}_i\cdot\mathbf{e}_j = \gamma_{i'i}\delta_{ij} = \gamma_{i'j},$$

or equivalently,

$$\mathbf{e}_{i'}\cdot\mathbf{e}_i = \gamma_{i'i}.$$

But $\mathbf{e}_i$ and $\mathbf{e}_{i'}$ are unit vectors, and hence

$$\mathbf{e}_{i'}\cdot\mathbf{e}_i = \cos(\widehat{\mathbf{e}_{i'}, \mathbf{e}_i}),$$

where $(\widehat{\mathbf{e}_{i'}, \mathbf{e}_i})$ denotes the angle between $\mathbf{e}_{i'}$ and $\mathbf{e}_i$. It follows that

$$\gamma_{i'i} = \cos(\widehat{\mathbf{e}_{i'}, \mathbf{e}_i}). \tag{2}$$

In just the same way, the vectors of the "old" basis $\mathbf{e}_1, \mathbf{e}_2, \mathbf{e}_3$ can be expressed as linear combinations of the vectors of the "new" basis $\mathbf{e}_{1'}, \mathbf{e}_{2'}, \mathbf{e}_{3'}$. Let $\gamma_{ii'}$ denote the coefficient of $\mathbf{e}_{i'}$ in the expansion of $\mathbf{e}_i$ with respect to the new basis vectors. Then the expansions of the old basis vectors with respect to the new basis vectors take the form

$$\begin{aligned}\mathbf{e}_1 &= \gamma_{11'}\mathbf{e}_{1'} + \gamma_{12'}\mathbf{e}_{2'} + \gamma_{13'}\mathbf{e}_{3'},\\ \mathbf{e}_2 &= \gamma_{21'}\mathbf{e}_{1'} + \gamma_{22'}\mathbf{e}_{2'} + \gamma_{23'}\mathbf{e}_{3'},\\ \mathbf{e}_3 &= \gamma_{31'}\mathbf{e}_{1'} + \gamma_{32'}\mathbf{e}_{2'} + \gamma_{33'}\mathbf{e}_{3'},\end{aligned}$$

or more concisely,

$$\mathbf{e}_i = \gamma_{ii'}\mathbf{e}_{i'}. \tag{3}$$

Taking the scalar product of each of the equations (3) with each of the vectors $\mathbf{e}_{j'}$, we get

$$\mathbf{e}_i\cdot\mathbf{e}_{j'} = \gamma_{ii'}\mathbf{e}_{i'}\cdot\mathbf{e}_{j'} = \gamma_{ii'}\delta_{i'j'} = \gamma_{ij'},$$

or equivalently,

$$\mathbf{e}_i\cdot\mathbf{e}_{i'} = \gamma_{ii'} = \cos\,(\widehat{\mathbf{e}_i, \mathbf{e}_{i'}}), \tag{4}$$

where $(\widehat{\mathbf{e}_i, \mathbf{e}_{i'}})$ denotes the angle between $\mathbf{e}_i$ and $\mathbf{e}_{i'}$. Since obviously $(\widehat{\mathbf{e}_i, \mathbf{e}_{i'}}) = (\widehat{\mathbf{e}_{i'}, \mathbf{e}_i})$, it follows from (2) and (4) that

$$\gamma_{i'i} = \gamma_{ii'}. \tag{5}$$

The numbers $\gamma_{i'i}$ can be written in the form of an array or *matrix*

$$\Gamma = \begin{pmatrix}\gamma_{1'1} & \gamma_{1'2} & \gamma_{1'3}\\ \gamma_{2'1} & \gamma_{2'2} & \gamma_{2'3}\\ \gamma_{3'1} & \gamma_{3'2} & \gamma_{3'3}\end{pmatrix}. \tag{6}$$

A matrix like (6), with the number of rows equal to the number of columns, is called a *square matrix*, and the number of rows (or columns) is called the *order* of the matrix. Thus Γ is a square matrix of order three, called the *matrix of the transformation from the old basis to the new basis*. Similarly, the numbers $\gamma_{ii'}$ form a matrix

$$\Gamma^{-1} = \begin{pmatrix}\gamma_{11'} & \gamma_{12'} & \gamma_{13'}\\ \gamma_{21'} & \gamma_{22'} & \gamma_{23'}\\ \gamma_{31'} & \gamma_{32'} & \gamma_{33'}\end{pmatrix},$$

called the *matrix of the transformation from the new basis to the old basis* (the notation Γ^{-1} shows that this is the matrix of the *inverse* transformation). The matrices Γ and Γ^{-1} can be written more concisely in the form

$$\Gamma = (\gamma_{i'i}), \qquad \Gamma^{-1} = (\gamma_{ii'}).$$

Formula (5) shows that the matrix Γ^{-1} is obtained from the matrix Γ by interchanging rows and columns in Γ. Moreover the elements of the two matrices satisfy the relations

$$\begin{aligned}\gamma_{i'k}\gamma_{j'k} &= \gamma_{ki'}\gamma_{kj'} = \delta_{i'j'},\\ \gamma_{ik'}\gamma_{jk'} &= \gamma_{k'i}\gamma_{k'j} = \delta_{ij}.\end{aligned} \tag{7}$$

In fact,

$$\gamma_{i'k}\gamma_{j'k} = \gamma_{i'1}\gamma_{j'1} + \gamma_{i'2}\gamma_{j'2} + \gamma_{i'3}\gamma_{j'3} = \mathbf{e}_{i'}\cdot\mathbf{e}_{j'} = \delta_{i'j'},$$

and similarly for the second formula. The relations (7) show that for either of the matrices Γ and Γ^{-1} *the sum of the products of the elements of any row (or column) with the corresponding elements of any other row (or column) equals zero, while the sum of the squares of the elements of any row (or column) equals unity.* A matrix whose elements satisfy these conditions is said to be *orthogonal.* Thus we have shown that *the transformation from one orthonormal basis to another in L_3 is described by an orthogonal matrix.* Conversely, let $\Gamma = (\gamma_{i'i})$ be any orthogonal matrix. Then, by (7), the vectors $\mathbf{e}_{i'}$ defined by (1) form a set of orthogonal unit vectors. It follows that *every orthogonal matrix is the matrix of the transformation from one orthonormal basis to another.*

Let $|\Gamma|$ denote the determinant of the matrix Γ, so that

$$|\Gamma| = \begin{vmatrix} \gamma_{1'1} & \gamma_{1'2} & \gamma_{1'3} \\ \gamma_{2'1} & \gamma_{2'2} & \gamma_{2'3} \\ \gamma_{3'1} & \gamma_{3'2} & \gamma_{3'3} \end{vmatrix}.$$

Then, since the rows of $|\Gamma|$ are made up of the components of the vectors $\mathbf{e}_{1'}, \mathbf{e}_{2'}, \mathbf{e}_{3'}$ with respect to the basis $\mathbf{e}_1, \mathbf{e}_2, \mathbf{e}_3$, it follows from formula (5), p. 19, that

$$|\Gamma| = \epsilon(\mathbf{e}_{1'}, \mathbf{e}_{2'}, \mathbf{e}_{3'}),$$

where the scalar triple product on the right is of absolute value 1, being equal to the volume of the cube constructed on the vectors $\mathbf{e}_{1'}, \mathbf{e}_{2'}, \mathbf{e}_{3'}$. Hence *the determinant of any orthogonal matrix equals* ± 1, where the plus sign is chosen if the bases $\mathbf{e}_1, \mathbf{e}_2, \mathbf{e}_3$ and $\mathbf{e}_{1'}, \mathbf{e}_{2'}, \mathbf{e}_{3'}$ have the same handedness and the minus sign otherwise (cf. p. 17). In the first case, the basis $\mathbf{e}_1, \mathbf{e}_2, \mathbf{e}_3$ can be brought into coincidence with the basis $\mathbf{e}_{1'}, \mathbf{e}_{2'}, \mathbf{e}_{3'}$ by making a rotation about the point O, while in the second case a rotation alone will not suffice and in fact we must also make a reflection of the basis $\mathbf{e}_1, \mathbf{e}_2, \mathbf{e}_3$ in some plane through O.

Example. In the plane a transformation from one orthonormal basis to another is either a pure rotation through some angle θ (in the counterclockwise direction, say) about an origin O, or else such a rotation followed by reflection in some line through O. In the first case, the formulas for the transformation of the basis are of the form†

$$\mathbf{e}_{1'} = \mathbf{e}_1 \cos\theta + \mathbf{e}_2 \sin\theta,$$
$$\mathbf{e}_{2'} = -\mathbf{e}_1 \sin\theta + \mathbf{e}_2 \cos\theta,$$

so that Γ, the matrix of the transformation, becomes

$$\Gamma = \begin{pmatrix} \cos\theta & \sin\theta \\ -\sin\theta & \cos\theta \end{pmatrix},$$

† See Example 6, p. 72.

with determinant 1. In the second case, the transformation formulas are of the form

$$\mathbf{e}_{1'} = \mathbf{e}_1 \cos\theta + \mathbf{e}_2 \sin\theta,$$
$$\mathbf{e}_{2'} = \mathbf{e}_1 \sin\theta - \mathbf{e}_2 \cos\theta,$$

so that Γ becomes

$$\Gamma = \begin{pmatrix} \cos\theta & \sin\theta \\ \sin\theta & -\cos\theta \end{pmatrix},$$

with determinant -1.

6.2. Consider any spatial vector $\mathbf{x}$. The vector $\mathbf{x}$ represents some geometrical or physical object, specified both in magnitude and direction, e.g., a force, velocity, acceleration, or electric field intensity. This "real" object does not depend on the coordinate system in which it is considered, and hence any operations or calculations *directly* involving vectors must always have a physical interpretation. However, together with direct calculations on vectors, a great role is played in geometry and its applications by the coordinate (or component) method, whose use permits us to study geometrical objects indirectly, by well-developed methods of both algebra (in analytic geometry) and analysis (in differential geometry). These methods allow us to obtain a number of results quite simply, whose direct proof would sometimes be very formidable or even impossible. On the other hand, in applying the coordinate method, we associate with the vector $\mathbf{x}$ its components x_1, x_2 and x_3, numbers which depend not only on the vector $\mathbf{x}$ itself, but also on the particular coordinate system (orthonormal basis) under consideration. Orthonormal bases can be chosen in many ways. For example, having chosen one basis, we can get many other bases by rotating the original basis about the origin of coordinates. Thus in applying the coordinate method we deal with data which reflect not only the geometrical situation but also the arbitrariness implicit in the selection of a coordinate system. For example, the very components of a vector depend on the coordinate system, while the sum of the squares of these components (which, as we know, gives the square of the length of the vector) ought not to depend on the choice of the coordinate system, and in fact, as we will see in a moment, this quantity turns out to be the same in all orthonormal bases. The properties of geometrical or physical objects which do not depend on the choice of the coordinate system (in which the given object is considered) are called *invariant* properties, and it is just such properties which are of primary interest.

This preliminary discussion leads to the

FUNDAMENTAL PROBLEM OF TENSOR CALCULUS. *How does one formulate propositions involving geometrical and physical objects in a way free from the influence of the underlying arbitrarily chosen coordinate system?*

As a first step towards the solution of this problem, we now examine how the components of a vector $\mathbf{x}$ transform in going from one orthonormal basis $\mathbf{e}_1, \mathbf{e}_2, \mathbf{e}_3$ with origin O to another orthonormal basis $\mathbf{e}'_1, \mathbf{e}'_2, \mathbf{e}'_3$ with the same origin. Let

$$\mathbf{x} = x_i\mathbf{e}_i, \qquad \mathbf{x} = x_{i'}\mathbf{e}_{i'}$$

be the expansions of $\mathbf{x}$ with respect to each of these bases. Since these are expansions of one and the same vector, we can equate the right-hand sides, obtaining

$$x_i\mathbf{e}_i = x_{i'}\mathbf{e}_{i'}. \tag{8}$$

Using (3) to replace the vectors $\mathbf{e}_i$ by their expansions relative to the basis $\mathbf{e}_{1'}, \mathbf{e}_{2'}, \mathbf{e}_{3'}$, we get

$$x_i\gamma_{ii'}\mathbf{e}_{i'} = x_{i'}\mathbf{e}_{i'},$$

which, because of the linear independence of the vectors $\mathbf{e}_{i'}$, implies

$$x_{i'} = x_i\gamma_{ii'},$$

or equivalently,

$$x_{i'} = \gamma_{i'i}x_i, \tag{9}$$

where we use the fact that $\gamma_{ii'} = \gamma_{i'i}$. *Formula* (9) *expresses the new components of the vector* $\mathbf{x}$ *in terms of its old components*. Alternatively, if we use (1) to replace the vectors $\mathbf{e}_{i'}$ in (8) by their expansions relative to the basis $\mathbf{e}_1, \mathbf{e}_2, \mathbf{e}_3$, we get the formula

$$x_i = \gamma_{ii'}x_{i'}, \tag{10}$$

expressing the old components of the vector $\mathbf{x}$ *in terms of its new components.* It should be noted that (10) can be obtained from (9) by multiplying both sides of (9) by $\gamma_{ji'}$, summing over i', and then using (7).

Next we examine which of the considerations of this chapter are of an invariant nature, i.e., are independent of the choice of the coordinate system, beginning with the case of the scalar product. The scalar product in the three-dimensional space L_3 was defined purely geometrically on p. 12, and hence there can be no doubt about its invariance. We now prove this invariance once again, starting from formula (1), p. 13, which expresses the scalar product of two vectors $\mathbf{x}$ and $\mathbf{y}$ in terms of the components of $\mathbf{x}$ and $\mathbf{y}$ relative to some orthonormal basis. It is important to do this, since the scalar product of two vectors $\mathbf{x}$ and $\mathbf{y}$ in the n-dimensional space L_n is *defined* as the sum of the components of $\mathbf{x}$ and $\mathbf{y}$ relative to some orthonormal basis (see Sec. 4, Prob. 5), so that in this case the required invariance cannot be of a geometrical character and must be proved *analytically* (in fact by precisely the same argument as we will now use to carry out the proof in the three-dimensional case).

Thus let $\mathbf{x}$ and $\mathbf{y}$ be two vectors in L_3, with components x_i, y_i relative to an orthonormal basis $\mathbf{e}_1, \mathbf{e}_2, \mathbf{e}_3$ and components $x_{i'}, y_{i'}$ relative to another orthonormal basis $\mathbf{e}_{1'}, \mathbf{e}_{2'}, \mathbf{e}_{3'}$. Then the scalar product $\mathbf{x}\cdot\mathbf{y}$ can be written either as x_iy_i or as $x_{i'}y_{i'}$. To prove the identity of these two expressions (and hence the invariance of the scalar product), we need only use (7) and (9) to deduce that

$$x_{i'}y_{i'} = \gamma_{i'i}x_i\gamma_{i'j}y_j = \delta_{ij}x_iy_j = x_iy_i.$$

The invariance of the formula for the scalar product immediately implies the invariance of the formulas for the length of a vector and for the cosine of the angle between two vectors, since these quantities are expressed in terms of the scalar product (see p. 13).

Before proving the invariance of the formulas expressing the vector product of two vectors and the scalar triple product of three vectors in terms of their components, we study the behavior of the components of the antisymmetric Kronecker symbol (see p. 17) under transformation to a new basis. In the new basis we have

$$\epsilon_{i'j'k'} = (\mathbf{e}_{i'}, \mathbf{e}_{j'}, \mathbf{e}_{k'})$$

(see p. 19). Since

$$\mathbf{e}_{i'} = \gamma_{i'i}\mathbf{e}_i, \qquad \mathbf{e}_{j'} = \gamma_{j'j}\mathbf{e}_j, \qquad \mathbf{e}_{k'} = \gamma_{k'k}\mathbf{e}_k,$$

it follows that

$$\epsilon_{i'j'k'} = \gamma_{i'i}\gamma_{j'j}\gamma_{k'k}\epsilon_{ijk}.$$

Remark. In particular,

$$\epsilon_{1'2'3'} = \gamma_{1'i}\gamma_{2'j}\gamma_{3'k}\epsilon_{ijk}, \tag{11}$$

where there are only six nonzero terms in the right-hand side, so that, in expanded form, (11) becomes

$$\epsilon_{1'2'3'} = (\gamma_{1'1}\gamma_{2'2}\gamma_{3'3} + \gamma_{1'2}\gamma_{2'3}\gamma_{3'1} + \gamma_{1'3}\gamma_{2'1}\gamma_{3'2}$$
$$- \gamma_{1'2}\gamma_{2'1}\gamma_{3'3} - \gamma_{1'3}\gamma_{2'2}\gamma_{3'1} - \gamma_{1'1}\gamma_{2'3}\gamma_{3'2})\epsilon_{123}.$$

The quantity in parentheses is just the determinant of the transformation matrix (6), and hence

$$\epsilon_{1'2'3'} = \epsilon_{123} \det \Gamma. \tag{12}$$

Letting ϵ' denote the value of the quantity ϵ relative to the new basis, we can write (12) as

$$\epsilon' = \epsilon \det \Gamma \tag{13}$$

(recall from p. 17 that $\epsilon_{123} = \epsilon$). Formula (13) shows that if the handedness of the basis is changed, then ϵ does not change, while if the handedness of the basis is reversed, then ϵ changes sign, in keeping with the original definition of the quantity ϵ on p. 17.

Now let

$$\mathbf{z} = \mathbf{x} \times \mathbf{y},$$

so that

$$z_k = \epsilon_{ijk}x_iy_j, \tag{14}$$

while

$$z_{k'} = \epsilon_{i'j'k'}x_{i'}y_{j'} \tag{14'}$$

in the new basis. To prove the invariance of the expressions for the components of **z**, i.e., to show that formula (14) goes into formula (14′) under a transformation of basis, we first substitute the expansions

$$x_i = \gamma_{ii'}x_{i'}, \qquad y_j = \gamma_{jj'}y_{j'}, \qquad z_k = \gamma_{kk'}z_{k'}$$

into (14), obtaining

$$\gamma_{kk'}z_{k'} = \epsilon_{ijk}\gamma_{ii'}\gamma_{jj'}x_{i'}y_{j'}. \tag{15}$$

We then multiply (15) by $\gamma_{kl'}$ and sum over the index k. Since

$$\gamma_{kk'}\gamma_{kl'} = \delta_{k'l'},$$

this gives

$$z_{l'} = \epsilon_{ijk}\gamma_{ii'}\gamma_{jj'}\gamma_{kl'}x_{i'}y_{j'},$$

which coincides with (14′), since

$$\epsilon_{ijk}\gamma_{ii'}\gamma_{jj'}\gamma_{kl'} = \gamma_{i'i}\gamma_{j'j}\gamma_{l'k}\epsilon_{ijk} = \epsilon_{i'j'l'},$$

as shown above.

In just the same way, we can prove that formula (4), p. 19 for the scalar triple product of three vectors **x**, **y**, and **z** is invariant under a transformation of basis, i.e., that

$$\epsilon_{ijk}x_iy_jz_k = \epsilon_{i'j'k'}x_{i'}y_{j'}z_{k'}.$$

However, the invariance of the expression for the scalar triple product follows even more simply from the formula

$$(\mathbf{x}, \mathbf{y}, \mathbf{z}) = (\mathbf{x} \times \mathbf{y})\cdot\mathbf{z}$$

and the fact that scalar and vector products are given by invariant expressions, as just proved

PROBLEMS

1. Let $\mathbf{e}_1, \mathbf{e}_2$ and $\mathbf{e}_{1'}, \mathbf{e}_{2'}$ be two orthonormal bases in the space L_2. Express the vectors of one basis in terms of those of the other basis and write the formulas for the transformation of an arbitrary vector in going from one basis to another if

a) The vectors of the second basis are obtained from those of the first basis by rotation through the angle α (in the counterclockwise direction) followed by relabelling of the basis vectors;

b) $\mathbf{e}_{1'} = -\mathbf{e}_1, \mathbf{e}_{2'} = \mathbf{e}_2$.

2. Write the matrix Γ of the transformation from one orthonormal basis $\mathbf{e}_1, \mathbf{e}_2, \mathbf{e}_3$ in the space L_3 to another orthonormal basis $\mathbf{e}_{1'}, \mathbf{e}_{2'}, \mathbf{e}_{3'}$ if

a) $\mathbf{e}_{1'} = \mathbf{e}_2, \mathbf{e}_{2'} = \mathbf{e}_1, \mathbf{e}_{3'} = \mathbf{e}_3$;

b) $\mathbf{e}_{1'} = \mathbf{e}_3, \mathbf{e}_{2'} = \mathbf{e}_1, \mathbf{e}_{3'} = \mathbf{e}_2$.

3. How does the matrix of the transformation from one basis to another change if

a) Two vectors of the first basis are interchanged;

b) Two vectors of the second basis are interchanged;

c) The vectors of both bases are written in reverse order?

4. Given two right-handed orthonormal bases $\mathbf{e}_1, \mathbf{e}_2, \mathbf{e}_3$ and $\mathbf{e}_{1'}, \mathbf{e}_{2'}, \mathbf{e}_{3'}$ in the space L_3, suppose the position of the second basis with respect to the first basis is specified by the three *Eulerian angles*, namely

a) The angle θ between the vectors $\mathbf{e}_3$ and $\mathbf{e}_{3'}$, given by the formula

$$\cos\theta = \mathbf{e}_3 \cdot \mathbf{e}_{3'};$$

b) The angle φ between the vectors $\mathbf{e}_1$ and $\mathbf{u}$, where $\mathbf{u}$ is a unit vector lying on the *line of nodes*, i.e., the intersection of the plane determined by $\mathbf{e}_1$, $\mathbf{e}_2$ and the plane determined by $\mathbf{e}_{1'}$, $\mathbf{e}_{2'}$, where $\mathbf{u}$, $\mathbf{e}_3$ and $\mathbf{e}_{3'}$ form a right-handed triple;

c) The angle ψ between the vectors $\mathbf{u}$ and $\mathbf{e}_{1'}$.

Express the vectors of the second basis in terms of those of the first basis, using the angles θ, φ and ψ.

5. Prove that the matrices

$$\Gamma_1 = \begin{pmatrix} \frac{2}{3} & \frac{2}{3} & -\frac{1}{3} \\ \frac{2}{3} & -\frac{1}{3} & \frac{2}{3} \\ -\frac{1}{3} & \frac{2}{3} & \frac{2}{3} \end{pmatrix}, \qquad \Gamma_2 = \begin{pmatrix} \frac{1}{2} & \frac{1}{2} & \frac{\sqrt{2}}{2} \\ \frac{1}{2} & \frac{1}{2} & -\frac{\sqrt{2}}{2} \\ \frac{\sqrt{2}}{2} & -\frac{\sqrt{2}}{2} & 0 \end{pmatrix}$$

are orthogonal.

6. Every formula involving a change of basis in the three-dimensional space L_3 is valid for the n-dimensional space L_n, provided only that we let the indices i, j, k, i', j', k', etc. take all values from 1 to n rather than just the values 1, 2, 3. Suppose a vector $\mathbf{x} \in L_n$ has components $x_1, x_2, \ldots, x_n$ with respect to an orthonormal basis $\mathbf{e}_1, \mathbf{e}_2, \ldots, \mathbf{e}_n$. What choice of a new basis in L_n makes the components of the vector $\mathbf{x}$ equal to $0, 0, \ldots, |\mathbf{x}|$?

7. Let $\mathbf{e}_1, \mathbf{e}_2, \ldots, \mathbf{e}_n$ be a basis in the space L_n, and let L_k be a nontrivial subspace of L_n of dimension k. Prove that L_k can be specified as the set of all vectors $\mathbf{x} \in L_n$ whose components $x_1, x_2, \ldots, x_n$ relative to the basis $\mathbf{e}_1, \mathbf{e}_2, \ldots, \mathbf{e}_n$ satisfy a system of equations of the form

$$a_{ij}x_j = 0$$

$(i = 1, 2, \ldots, m \le n)$.

8. In the space of all polynomials of degree not exceeding n,† write the matrix of the transformation from the basis $1, t, \ldots, t^n$ to the basis $1, t - a, \ldots, (t - a)^n$. Write formulas for the transformation of the coefficients of an arbitrary polynomial under such a change of basis.

7. Topics in Analytic Geometry

We now consider some topics in analytic geometry, with the aim of recalling a number of facts that will be needed later. At the same time, we will use this occasion to write the relevant equations in concise notation (the notation in which they will be used later).

† See Sec. 1, Example 6, Sec. 2, Prob. 5 and Sec. 3, Prob. 2.

Let O be a fixed point of ordinary (Euclidean) space. Then every spatial point P can be assigned a vector $\overrightarrow{OP} = \mathbf{x}$, called the *radius vector* of P. The position of P is uniquely determined once we know its radius vector $\mathbf{x}$ (with respect to the given origin O). In other words, given an origin O, we can establish a one-to-one correspondence between the vectors of the linear space L_3 (equipped with the usual scalar product) and the points of ordinary space. The components x_1, x_2, x_3 of the vector $\mathbf{x}$ with respect to an orthonormal basis $\mathbf{e}_1, \mathbf{e}_2, \mathbf{e}_3$ are just the coordinates of the point P with respect to a rectangular coordinate system whose origin coincides with O and whose axes are directed along the vectors $\mathbf{e}_1, \mathbf{e}_2, \mathbf{e}_3$.

Naturally, this correspondence between points and vectors pertains to a given *fixed* origin O. If we go from O to a new origin O', the radius vector of every point P changes correspondingly. Let $\overrightarrow{O'P} = \mathbf{x}'$ be the new radius vector of the point P, with respect to the new origin O'. Then the relation between the old and new radius vectors is given by

$$\mathbf{x} = \mathbf{x}' + \mathbf{p} \tag{1}$$

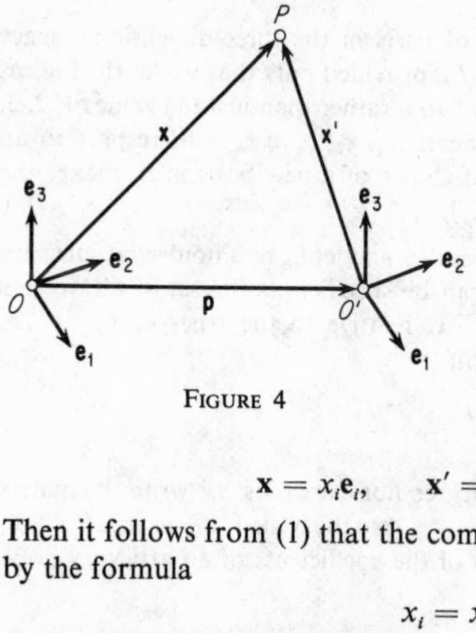

FIGURE 4

(see Figure 4). Consider two rectangular coordinate systems with origins at the points O and O', respectively, whose axes are parallel and determined by the orthonormal vectors $\mathbf{e}_1, \mathbf{e}_2, \mathbf{e}_3$. Then the coordinates of P with respect to the first system are just the components of the vector $\mathbf{x}$, while the coordinates of P with respect to the second system are just the components of the vector $\mathbf{x}'$. Let the expansions of the vectors $\mathbf{x}, \mathbf{x}'$ and $\mathbf{p}$ with respect to the basis $\mathbf{e}_1, \mathbf{e}_2, \mathbf{e}_3$ be

$$\mathbf{x} = x_i\mathbf{e}_i, \quad \mathbf{x}' = x_i'\mathbf{e}_i, \quad \mathbf{p} = p_i\mathbf{e}_i.$$

Then it follows from (1) that the components of these vectors are connected by the formula

$$x_i = x_i' + p_i, \tag{1'}$$

which shows how the coordinates of the point P transforms when the coordinate system is shifted parallel to itself.

Remark 1. Note that shifting a coordinate system parallel to itself has no effect on vectors, since such shifts do not change the basis vectors.

Remark 2. As already noted in the preceding section, all quantities and equations which have any geometric meaning must remain invariant (i.e., unchanged) under arbitrary transformations of a rectangular coordinate system. Since the components of vectors change only when the underlying orthonormal basis is changed and do not change under parallel shifts of the coordinate axes, every quantity depending on the components of vectors is invariant under parallel shifts of the coordinate system. Hence we need only test the invariance of such quantities under rotations of the coordinate system. By contrast, quantities which depend on the coordinates of points can change not only under rotations of the coordinate system, but also under parallel shifts of the coordinate system. Hence we must test the invariance of quantities depending on the coordinates of points under transformations of both types.

We now consider a number of concrete problems arising in three-dimensional analytic geometry.

7.1. Distance between two points. Division of a line segment in a given ratio. Let P and Q be two points in space, with radius vectors $\mathbf{x}$ and $\mathbf{y}$. Then $\overrightarrow{PQ} = \mathbf{y} - \mathbf{x}$, and the length of the segment PQ equals

$$\begin{aligned} |\overrightarrow{PQ}| = |\mathbf{y} - \mathbf{x}| &= \sqrt{\delta_{ij}(y_i - x_i)(y_j - x_j)} \\ &= \sqrt{(y_1 - x_1)^2 + (y_2 - x_2)^2 + (y_3 - x_3)^2}. \end{aligned} \tag{2}$$

The invariance of this expression under arbitrary (orthogonal) coordinate transformations follows from the fact that the distance between the points P and Q equals the length of the vector $\overrightarrow{PQ}$, which, as we have seen, is invariant under such transformations.

The point M dividing the segment PQ in the ratio λ, i.e., such that†

$$\frac{|PM|}{|MQ|} = \lambda,$$

is specified by the radius vector $\mathbf{z}$ such that

$$\mathbf{z} - \mathbf{x} = \lambda(\mathbf{y} - \mathbf{z}),$$

which implies

$$\mathbf{z} = \frac{\mathbf{x} + \lambda\mathbf{y}}{1 + \lambda}. \tag{3}$$

The components of this vector are related to those of the vectors $\mathbf{x}$ and $\mathbf{y}$ by the formula

$$z_i = \frac{x_i + \lambda y_i}{1 + \lambda}. \tag{3'}$$

Suppose we subject the coordinate system to a (parallel) shift defined by the vector $\mathbf{p}$. Then (3) becomes

$$z' = \frac{\mathbf{x}' + \lambda\mathbf{y}'}{1 + \lambda},$$

† $|PM|$ denotes the length of the segment PM, and similarly for $|MQ|$.

where

$$\mathbf{x} = \mathbf{x}' + \mathbf{p}, \qquad \mathbf{y} = \mathbf{y}' + \mathbf{p}, \qquad \mathbf{z} = \mathbf{z}' + \mathbf{p}.$$

It follows that (3) is invariant under shifts.

7.2. Equation of a plane. Let Π be a plane in space, and let $\mathbf{n}$ be a vector normal to Π. Suppose $\mathbf{n}$ has components a_i with respect to an orthonormal basis $\mathbf{e}_1, \mathbf{e}_2, \mathbf{e}_3$, so that

$$\mathbf{n} = a_i\mathbf{e}_i. \tag{4}$$

Let $\mathbf{x}_0$ be the radius vector of a fixed point $P_0 \in \Pi$, with components x_i^0, and let $\mathbf{x}$ be the radius vector of an arbitrary vector $P \in \Pi$, with components x_i. Then clearly

$$\mathbf{x}_0 = x_i^0\mathbf{e}_i, \qquad \mathbf{x} = x_i\mathbf{e}_i,$$

and hence

$$\overrightarrow{P_0P} = \mathbf{x} - \mathbf{x}_0 = (x_i - x_i^0)\mathbf{e}_i. \tag{5}$$

Since the vectors $\overrightarrow{P_0P}$ and $\mathbf{n}$ are perpendicular, we have

$$\overrightarrow{P_0P}\cdot\mathbf{n} = 0,$$

or

$$(\mathbf{x} - \mathbf{x}_0)\cdot\mathbf{n} = 0, \tag{6}$$

a result known as the *vector form* of the equation of Π.

Using (4), (5) and the expression for the scalar product of two vectors in terms of the components of the vectors, we get

$$a_i(x_i - x_i^0) = 0,$$

or

$$a_ix_i + b = 0 \tag{6'}$$

after denoting $-a_ix_i^0$ by b. Equation (6′) can be written in the form

$$\mathbf{n}\cdot\mathbf{x} + b = 0.$$

If the plane Π passes through the origin of coordinates, then $b = 0$, and the equation of the plane becomes

$$a_ix_i = 0.$$

On the other hand, if Π does not pass through the origin, then $b \neq 0$ and we can divide all the terms of equation (6′) by b, obtaining

$$u_ix_i = 1$$

in terms of the numbers

$$u_i = -\frac{a_i}{b}.$$

7.3. Distance from a point to a plane. Suppose the plane Π has equation

$$a_ix_i + b = 0$$

in some orthonormal basis $\mathbf{e}_1, \mathbf{e}_2, \mathbf{e}_3$. Then we can write the unit normal $\mathbf{n}_0$ to Π in the form

$$\mathbf{n}_0 = \frac{a_i\mathbf{e}_i}{\sqrt{a_ia_i}}.$$

Let P_0 be any point of space, with coordinates x_i^0, and let P be any point of Π, with coordinates x_i. Then the distance δ from the point P_0 to the plane Π can be written in the form

$$\delta = |\mathrm{Pr}_{\mathbf{n}_0}\overrightarrow{PP_0}| = |\mathrm{Pr}_{\mathbf{n}_0}(x_i^0 - x_i)\mathbf{e}_i|$$
$$= \frac{|(x_i^0 - x_i)a_i|}{\sqrt{a_ia_i}} = \frac{|a_ix_i^0 + b|}{\sqrt{a_ia_i}}$$

(see p. 14). In particular, the distance δ_0 from the origin $O = (0, 0, 0)$ to the plane Π equals

$$\delta_0 = \frac{|b|}{\sqrt{a_ia_i}}.$$

7.4. Equation of a straight line in space. Let l be the line in space with the direction specified by the vector $\mathbf{a} = a_i\mathbf{e}_i$ which goes through the point P_0 with radius vector

$$\mathbf{r}_0 = x_i^0\mathbf{e}_i$$

in some orthonormal basis $\mathbf{e}_1, \mathbf{e}_2, \mathbf{e}_3$. Let P be an arbitrary point of l, with radius vector

$$\mathbf{r} = x_i\mathbf{e}_i.$$

Since the vectors

$$\overrightarrow{P_0P} = \mathbf{r} - \mathbf{r}_0 = (x_i - x_i^0)\mathbf{e}_i,$$

are collinear, we have

$$\mathbf{r} - \mathbf{r}_0 = \lambda\mathbf{a},$$

where λ is a parameter which can take arbitrary real values, or equivalently,

$$\mathbf{r} = \mathbf{r}_0 + \lambda\mathbf{a} \tag{7}$$

(the *vector form* of the equation of l). In coordinate form, (7) becomes

$$x_i = x_i^0 + \lambda a \qquad (i = 1, 2, 3) \tag{7'}$$

(the *parametric equations* of l).

7.5. The straight line as the intersection of two planes. Let l be the line of intersection of two planes Π_1 and Π_2. Then l is determined by the system of two equations

$$a_i^{(1)}x_i + b^{(1)} = 0, \qquad a_i^{(2)}x_i + b^{(2)} = 0, \tag{8}$$

where $a_i^{(1)}$ and $a_i^{(2)}$ are components of normal vectors $\mathbf{n}_1$ and $\mathbf{n}_2$ to the planes Π_1 and Π_2, respectively. To go from (8) to (7′), we must find a point P_0 on

l and a vector $\mathbf{a}$ parallel to l. Being perpendicular to the vectors $\mathbf{n}_1$ and $\mathbf{n}_2$, the vector $\mathbf{a}$ can be chosen as the vector product of $\mathbf{n}_1$ and $\mathbf{n}_2$:

$$\mathbf{a} = \mathbf{n}_1 \times \mathbf{n}_2 = \epsilon_{ijk} a_i^{(1)} a_j^{(2)} \mathbf{e}_k.$$

To find P_0 we need only fix one of the coordinates x_i and then solve the system (8) for the other two coordinates (one coordinate must be fixed if the system is to have a solution). Let x_i^0 be the coordinates of the point P_0 which we have found in this way. Then the parametric equations of l can be written in the form

$$x_k = x_k^0 + \lambda \epsilon_{ijk} a_i^{(1)} a_j^{(2)} \qquad (k = 1, 2, 3).$$

7.6. General equation of a second-degree curve in the plane. The general equation of a second-degree curve relative to some rectangular coordinate system in the plane is given by

$$Ax^2 + 2Bxy + Cy^2 + 2Dx + 2Ey + F = 0. \tag{9}$$

Let the coordinates x and y be denoted by x_1 and x_2. Moreover, let a_{ij} denote the coefficient of the product $x_i x_j$, let a_i denote the coefficient of x_i, and let a denote the constant term. Then (9) can be rewritten in the concise form

$$a_{ij} x_i x_j + 2a_i x_i + a = 0, \qquad (10),$$

where $a_{ij} = a_{ji}$. Note that in the first term summation takes place over both indices i and j. In fact, writing the first term out in detail, we get

$$\begin{aligned} a_{ij} x_i x_j &= a_{11} x_1^2 + a_{12} x_1 x_2 + a_{21} x_2 x_1 + a_{22} x_2^2 \\ &= a_{11} x_1^2 + 2a_{12} x_1 x_2 + a_{22} x_2^2. \end{aligned}$$

Hence, when written out in full, (10) becomes

$$a_{11} x_1^2 + 2a_{12} x_1 x_2 + a_{22} x_2^2 + 2a_1 x_1 + 2a_2 x_2 + a = 0, \tag{10'}$$

which coincides with (9). The condition

$$a_i = 0$$

means that the second-order curve is central, with the origin of coordinates as its center of symmetry (why?), while the conditions

$$a_i = 0, \qquad a = 0$$

mean that the curve degenerates into two intersecting (or coincident) lines passing through the origin.

7.7. General equation of a second-degree surface. The general equation of a second-degree (or quadric) surface relative to some rectangular coordinate system in space is given by

$$Ax^2 + By^2 + Cz^2 + 2Dxy + 2Exz + 2Fyz + 2Gx + 2Hy + 2Kz + L = 0. \tag{11}$$

Using notation analogous to that just introduced in the case of the second-degree curve, we can write (11) concisely as

$$a_{ij}x_ix_j + 2a_ix_i + a = 0, \tag{12}$$

where $a_{ij} = a_{ji}$. Note that (10) and (12) are identical, except for the fact that the indices of summation take the values 1, 2 in (10) and the values 1, 2, 3 in (12). As before, the condition

$$a_i = 0$$

means that the quadric surface is central, with the origin of coordinates as its center of symmetry, while the conditions

$$a_i = 0, \qquad a = 0$$

means that the surface is a cone with its center at the origin which, in particular, may degenerate into two intersecting (or coincident) planes passing through the origin.

7.8. Determination of the center of a second-degree curve or surface. We can often solve a problem involving a second-degree curve and the analogous problem involving a second-degree surface *simultaneously*, exploiting the fact that the curve and the surface both have the same equation (10) or (12) in concise notation (provided, of course, that we bear in mind that the indices of summation take two values for the curve and three values for the surface). Consider, for example, the problem of determining the center of a second-degree curve or surface, starting from the common equation (10) or (12). Suppose this equation pertains to a rectangular coordinate system with origin O, and suppose we shift the origin of the coordinate system to the *center* of the curve or surface. Let $\mathbf{p}$ be the radius vector of the new origin O' relative to the old origin O, i.e.,

$$\mathbf{p} = \overrightarrow{OO'} = p_i\mathbf{e}_i.$$

Then the old coordinates x_i and the new coordinates x'_i of a variable point P are related by formula (1′):

$$x_i = x'_i + p_i.$$

Substituting these values of x_i into equation (10) or (12), we find that the equation takes the form

$$a_{ij}(x'_i + p_i)(x'_j + p_j) + 2a_i(x'_i + p_i) + a = 0$$

or

$$a_{ij}x'_ix'_j + a_{ij}x'_ip_j + a_{ij}x'_jp_i + a_{ij}p_ip_j + 2a_ix'_i + 2a_ip_i + a = 0$$

in the new coordinate system. Interchanging the indices i and j in the third term and noting that $a_{ij} = a_{ji}$, we get

$$a_{ij}x'_ix'_j + 2(a_{ij}p_j + a_i)x'_i + a_{ij}p_ip_j + 2a_ip_i + a = 0.$$

Since the new origin is at the center of the curve or surface, we must have

$$a_{ij}p_j = -a_i. \tag{13}$$

Thus the center has coordinates p_j satisfying the system (13), and in fact a center exists if and only if the system (13) has a solution, i.e., if and only if its determinant (of order two for a curve or three for a surface) is nonzero.

PROBLEMS

1. Write equations in both vector and coordinate form for the plane

a) Passing through two given intersecting lines

$$\mathbf{x} = \mathbf{x}_1 + \lambda\mathbf{a}, \qquad \mathbf{x} = \mathbf{x}_1 + \mu\mathbf{b};$$

b) Passing through the line $\mathbf{x} = \mathbf{x}_1 + \lambda\mathbf{a}$ and the point P_0 with radius vector $\mathbf{x}_0$.

2. Give necessary and sufficient conditions for intersection, parallelism or coincidence of the two planes

$$a_i^{(1)}x_i + b^{(1)} = 0, \qquad a_i^{(2)}x_i + b^{(2)} = 0.$$

3. Find the distance between the two parallel planes

$$a_ix_i + b = 0, \qquad a_ix_i + b' = 0. \tag{14}$$

4. Write the equation of the plane parallel to the planes (14) lying midway between them.

5. Write the equation of the family of planes going through the line of intersection of the planes

$$a_i^{(1)}x_i + b^{(1)} = 0, \qquad a_i^{(2)}x_i + b^{(2)} = 0. \tag{15}$$

6. In the family of planes figuring in the preceding problem, find the plane

a) Passing through the point P_0 with coordinates $x_i^{(0)}$;

b) Perpendicular to the plane $a_i^{(3)}x_i + b^{(3)}$ in the family.

7. Find the angle between the planes (15). When are the planes orthogonal?

8. Write the equations of the planes in the family figuring in Prob. 5 which bisect the angles between the planes (15) determining the family.

9. Find the coordinates of the foot of the perpendicular dropped from the point P_0 with coordinates $x_i^{(0)}$ to the plane $a_ix_i + b = 0$.

10. Find the area of the triangle whose vertices A, B and C have coordinates x_i, y_i and z_i, respectively.

11. Find the volume of the tetrahedron whose vertices A, B, C and D have coordinates x_i, y_i, z_i and u_i, respectively.

12. Find the distance from the point P with radius vector $\mathbf{y}$ to the line $\mathbf{x} = \mathbf{x}_0 + \lambda\mathbf{a}$.

13. Find the distance between the two parallel lines

$$\mathbf{x} = \mathbf{x}_1 + \lambda\mathbf{a}, \qquad \mathbf{x} = \mathbf{x}_2 + \mu\mathbf{a}.$$

14. Given two skew lines

$$\mathbf{x} = \mathbf{x}_1 + \lambda\mathbf{a}_1, \qquad \mathbf{x} = \mathbf{x}_2 + \mu\mathbf{a}_2,$$

find

a) The angle between the lines;

b) The shortest distance between them.

2

MULTILINEAR FORMS AND TENSORS

8. Linear Forms

8.1. The basic operations of vector algebra were considered in the preceding chapter. We now turn to the study of the simplest scalar functions of one or several vector arguments.

Given a linear space L, by a *scalar function* $\varphi = \varphi(\mathbf{x})$ defined on L we mean a rule associating a number φ with each vector $\mathbf{x} \in L$. We call φ a *linear function* (of $\mathbf{x}$) or a *linear form* (in $\mathbf{x}$) if

1) $\varphi(\mathbf{x} + \mathbf{y}) = \varphi(\mathbf{x}) + \varphi(\mathbf{y})$ for arbitrary vectors $\mathbf{x}$ and $\mathbf{y}$;
2) $\varphi(\lambda\mathbf{x}) = \lambda\varphi(\mathbf{x})$ for an arbitrary vector $\mathbf{x}$ and real number λ.

Example 1. Let $\mathbf{a}$ be a fixed vector and $\mathbf{x}$ a variable vector of the space L_3. Then the scalar product

$$\varphi(\mathbf{x}) = \mathbf{a}\cdot\mathbf{x}$$

is a linear form in $\mathbf{x}$, since

$$\mathbf{a}\cdot(\mathbf{x} + \mathbf{y}) = \mathbf{a}\cdot\mathbf{x} + \mathbf{a}\cdot\mathbf{y}, \qquad \mathbf{a}\cdot(\lambda\mathbf{x}) = \lambda\mathbf{a}\cdot\mathbf{x},$$

by the properties of the scalar product (see p. 12).

Example 2. In particular, let $\mathrm{Pr}_l\, \mathbf{x}$ be the projection of the vector $\mathbf{x}$ onto the (directed) line l, i.e., let

$$\mathrm{Pr}_l\, \mathbf{x} = \mathbf{e}_l\cdot\mathbf{x},$$

where $\mathbf{e}_l$ is a unit vector along l. Then $\mathrm{Pr}_l\, \mathbf{x}$ is a linear form in $\mathbf{x}$, since clearly

$$\mathrm{Pr}_l\, (\mathbf{x} + \mathbf{y}) = \mathrm{Pr}_l\, \mathbf{x} + \mathrm{Pr}_l\, \mathbf{y}, \qquad \mathrm{Pr}_l\, (\lambda\mathbf{x}) = \lambda\mathrm{Pr}_l\, \mathbf{x}.$$

Example 3. Since any component x_i of a vector $\mathbf{x} \in L_3$ with respect to an orthonormal basis $\mathbf{e}_1, \mathbf{e}_2, \mathbf{e}_3$ can be represented in the form

$$x_i = \mathbf{e}_i \cdot \mathbf{x}$$

(see p. 13), x_i is also a linear form in $\mathbf{x}$.

Example 4. Let $\mathbf{a}$ and $\mathbf{b}$ be two noncollinear vectors of the space L_3. Then the scalar triple product $(\mathbf{a}, \mathbf{b}, \mathbf{x})$ is a linear form in $\mathbf{x}$, since

$$(\mathbf{a}, \mathbf{b}, \mathbf{x} + \mathbf{y}) = (\mathbf{a}, \mathbf{b}, \mathbf{x}) + (\mathbf{a}, \mathbf{b}, \mathbf{y}), \qquad (\mathbf{a}, \mathbf{b}, \lambda\mathbf{x}) = \lambda(\mathbf{a}, \mathbf{b}, \mathbf{x}),$$

by the properties of the scalar triple product (see p. 18).

Next, given an orthonormal basis $\mathbf{e}_1, \mathbf{e}_2, \mathbf{e}_3$ in L_3, we find an expression for a linear form $\varphi(\mathbf{x})$ in terms of the components of $\mathbf{x}$ with respect to $\mathbf{e}_1$, $\mathbf{e}_2, \mathbf{e}_3$. Let

$$\mathbf{x} = x_i \mathbf{e}_i.$$

By the linearity of φ,

$$\varphi(\mathbf{x}) = \varphi(x_i \mathbf{e}_i) = x_i \varphi(\mathbf{e}_i),$$

so that, writing

$$a_i = \varphi(\mathbf{e}_i),$$

we have

$$\varphi(\mathbf{x}) = a_i x_i. \tag{1}$$

The expression (1) is a homogeneous polynomial of degree one in the variables x_i. The coefficients a_i in (1) obviously depend on the choice of basis.

8.2. We now examine how the coefficients of a linear form $\varphi = \varphi(\mathbf{x})$ transform in going from one orthonormal basis $\mathbf{e}_1, \mathbf{e}_2, \mathbf{e}_3$ to another orthonormal basis $\mathbf{e}_{1'}, \mathbf{e}_{2'}, \mathbf{e}_{3'}$. Under such a transformation, we have

$$\mathbf{e}_{i'} = \gamma_{i'i} \mathbf{e}_i,$$

where $\Gamma = (\gamma_{i'i})$ is the matrix of the transformation from the old basis to the new basis (see p. 23). In the new basis φ takes the form

$$\varphi = a_{i'} x_{i'},$$

where the $x_{i'}$ are the new components of the vector $\mathbf{x}$ and the coefficients $a_{i'}$ are given by

$$a_{i'} = \varphi(\mathbf{e}_{i'}) = \varphi(\gamma_{i'i} \mathbf{e}_i) = \gamma_{i'i} \varphi(\mathbf{e}_i) = \gamma_{i'i} a_i.$$

Hence the coefficients of the linear form φ transform according to the law

$$a_{i'} = \gamma_{i'i} a_i \tag{2}$$

in going from the old basis to the new basis. Comparing (2) with formula (9), p. 26, we see that the coefficients of a linear form transform in exactly the same way as the components of a vector in going over to the new basis. In other words, the coefficients a_i of a linear form φ are the components of some vector†

$$\mathbf{a} = a_i \mathbf{e}_i.$$

Thus formula (1) shows that the linear form $\varphi = \varphi(\mathbf{x})$ can always be written as the scalar product of a fixed vector $\mathbf{a}$ and a variable vector $\mathbf{x}$, i.e.,

$$\varphi = \varphi(\mathbf{x}) = \mathbf{a} \cdot \mathbf{x}.$$

Remark. To interpret the vector $\mathbf{a}$ geometrically, consider the *level surfaces* of the linear form φ, characterized by the equation $\varphi = c$ or

$$\mathbf{a} \cdot \mathbf{x} = c. \tag{3}$$

Clearly (3) is the equation of a family of parallel planes, each of which has $\mathbf{a}$ as a normal vector, i.e., $\mathbf{a}$ is a common normal to the planes making up the level surfaces of the form φ.

PROBLEMS

1. Which of the following scalar functions of a vector argument are linear forms:

a) The function

$$\varphi(\mathbf{x}) = c_i x_i,$$

where the x_i are the components of the vector $\mathbf{x}$ relative to some basis in the space L_n and the c_i are fixed numbers;

b) The function

$$\varphi(\mathbf{x}) = x_1^2,$$

where x_1 is the first component of $\mathbf{x}$ relative to some basis in L_n;

c) The function

$$\varphi(\mathbf{x}) = c;$$

d) The function

$$\varphi[f(t)] = f(t_0) \qquad (a < t_0 < b)$$

defined on the space $C[a, b]$ of all functions $f(t)$ continuous in the interval $[a, b]$ (cf. Sec. 1, Example 7);

e) The function

$$\varphi[f(t)] = \int_a^b c(t) f(t)\, dt,$$

where $c(t)$ is a fixed function and $f(t)$ a variable function in the space $C[a, b]$?

2. Write the linear function considered in Example 4 in the form $\varphi(\mathbf{x}) = \mathbf{a} \cdot \mathbf{x}$.

---

† Note that $\mathbf{a} = a_i \mathbf{e}_i = a_{i'} \mathbf{e}_{i'}$, as follows directly from

$$a_i \mathbf{e}_i = a_i \gamma_{ii'} \mathbf{e}_{i'} = \gamma_{i'i} a_i \mathbf{e}_{i'} = a_{i'} \mathbf{e}_{i'}.$$

9. Bilinear Forms

9.1. A scalar function $\varphi = \varphi(\mathbf{x}, \mathbf{y})$ of two vector arguments $\mathbf{x}$ and $\mathbf{y}$ is called a *bilinear function* or *bilinear form* if it is linear in both its arguments, i.e., if

1) $\varphi(\mathbf{x}_1 + \mathbf{x}_2, \mathbf{y}) = \varphi(\mathbf{x}_1, \mathbf{y}) + \varphi(\mathbf{x}_2, \mathbf{y})$;
2) $\varphi(\lambda\mathbf{x}, \mathbf{y}) = \lambda\varphi(\mathbf{x}, \mathbf{y})$;
3) $\varphi(\mathbf{x}, \mathbf{y}_1 + \mathbf{y}_2) = \varphi(\mathbf{x}, \mathbf{y}_1) + \varphi(\mathbf{x}, \mathbf{y}_2)$;
4) $\varphi(\mathbf{x}, \lambda\mathbf{y}) = \lambda\varphi(\mathbf{x}, \mathbf{y})$.

Example 1. The scalar product of two vectors $\mathbf{x}$ and $\mathbf{y}$ is a bilinear form since it clearly has all the above properties.

Example 2. Let $\mathbf{a}$ be a fixed vector and $\mathbf{x}, \mathbf{y}$ variable vectors of the space L_3. Then it is easy to see that the scalar triple product $(\mathbf{a}, \mathbf{x}, \mathbf{y})$ is a bilinear form (in $\mathbf{x}$ and $\mathbf{y}$).

Example 3. Let $\alpha(\mathbf{x})$ and $\beta(\mathbf{y})$ be linear forms in the variable vectors $\mathbf{x}$ and $\mathbf{y}$, respectively. Then the product

$$\varphi(\mathbf{x}, \mathbf{y}) = \alpha(\mathbf{x})\beta(\mathbf{y})$$

is a bilinear form, since

$$\begin{aligned}\varphi(\mathbf{x}_1 + \mathbf{x}_2, \mathbf{y}) &= \alpha(\mathbf{x}_1 + \mathbf{x}_2)\beta(\mathbf{y}) = \alpha(\mathbf{x}_1)\beta(\mathbf{y}) + \alpha(\mathbf{x}_2)\beta(\mathbf{y}) \\ &= \varphi(\mathbf{x}_1, \mathbf{y}) + \varphi(\mathbf{x}_2, \mathbf{y}), \\ \varphi(\lambda\mathbf{x}, \mathbf{y}) &= \alpha(\lambda\mathbf{x})\beta(\mathbf{y}) = \lambda\alpha(\mathbf{x})\beta(\mathbf{y}) = \lambda\varphi(\mathbf{x}, \mathbf{y}),\end{aligned}$$

and similarly for the second argument.

9.2. Next, given an orthonormal basis $\mathbf{e}_1, \mathbf{e}_2, \mathbf{e}_3$ in L_3, we find an expression for a bilinear form $\varphi(\mathbf{x}, \mathbf{y})$ in terms of the components of $\mathbf{x}$ and $\mathbf{y}$ with respect to $\mathbf{e}_1, \mathbf{e}_2, \mathbf{e}_3$. Let

$$\mathbf{x} = x_i\mathbf{e}_i, \quad \mathbf{y} = y_j\mathbf{e}_j.$$

By the linearity of φ in both its arguments,

$$\varphi(\mathbf{x}, \mathbf{y}) = \varphi(x_i\mathbf{e}_i, y_j\mathbf{e}_j) = x_i y_j \varphi(\mathbf{e}_i, \mathbf{e}_j),$$

so that, writing

$$a_{ij} = \varphi(\mathbf{e}_i, \mathbf{e}_j),$$

we have

$$\varphi(\mathbf{x}, \mathbf{y}) = a_{ij}x_i y_j,$$

or, in more detail,

$$\begin{aligned}\varphi(\mathbf{x}, \mathbf{y}) = a_{11}x_1y_1 + a_{12}x_1y_2 + a_{13}x_1y_3 + a_{21}x_2y_1 + a_{22}x_2y_2 \\ + a_{23}x_2y_3 + a_{31}x_3y_1 + a_{32}x_3y_2 + a_{33}x_3y_3.\end{aligned}$$

This expression is a homogeneous polynomial of degree two, linear in both sets of variables x_1, x_2, x_3 and y_1, y_2, y_3.

The coefficients of the bilinear form φ can be written in the form of an array

$$A = \begin{pmatrix} a_{11} & a_{12} & a_{13} \\ a_{21} & a_{22} & a_{23} \\ a_{31} & a_{32} & a_{33} \end{pmatrix},$$

i.e., as a square matrix of order three (see p. 23). The matrix A is called the *(coefficient) matrix of the bilinear form* φ. Thus, relative to a given basis $\mathbf{e}_1, \mathbf{e}_2, \mathbf{e}_3 \in L_3$, every bilinear form φ is characterized by a well-defined third-order matrix.

We now write the bilinear forms of Examples 1-3 in component form and find their matrices.

Example 1'. The bilinear form $\mathbf{x}\cdot\mathbf{y}$ becomes

$$\mathbf{x}\cdot\mathbf{y} = x_1y_1 + x_2y_2 + x_3y_3$$

in an orthonormal basis, and hence its matrix is just

$$\begin{pmatrix} 1 & 0 & 0 \\ 0 & 1 & 0 \\ 0 & 0 & 1 \end{pmatrix} = (\delta_{ij}).$$

Example 2'. Next consider the bilinear form $(\mathbf{a}, \mathbf{x}, \mathbf{y})$. Recalling the expression for the scalar triple product in component form (see p. 19), we have†

$$(\mathbf{a}, \mathbf{x}, \mathbf{y}) = \epsilon_{kij}a_kx_iy_j.$$

Hence the coefficient matrix of (a, x, y) takes the form

$$(\epsilon_{kij}a_k) = \epsilon\begin{pmatrix} 0 & a_3 & -a_2 \\ -a_3 & 0 & a_1 \\ a_2 & -a_1 & 0 \end{pmatrix}.$$

Example 3'. Relative to the orthonormal basis $\mathbf{e}_1, \mathbf{e}_2, \mathbf{e}_3$, the linear forms $\alpha(\mathbf{x})$ and $\beta(\mathbf{y})$ can be written as

$$\alpha(\mathbf{x}) = a_ix_i, \qquad \beta(\mathbf{y}) = b_jy_j$$

(see Sec. 8), so that the bilinear form $\varphi(\mathbf{x}, \mathbf{y}) = \alpha(\mathbf{x})\beta(\mathbf{y})$ becomes

$$\varphi(\mathbf{x}, \mathbf{y}) = a_ix_ib_jy_j = a_ib_jx_iy_j,$$

with matrix

$$(a_ib_j) = \begin{pmatrix} a_1b_1 & a_1b_2 & a_1b_3 \\ a_2b_1 & a_2b_2 & a_2b_3 \\ a_3b_1 & a_3b_2 & a_3b_3 \end{pmatrix}.$$

---

† Here we write the indices of summation somewhat differently than on p. 19.

9.3. Next we examine how the coefficients of a bilinear form $\varphi = \varphi(\mathbf{x}, \mathbf{y})$ transform under a change of basis. Relative to a new orthonormal basis $\mathbf{e}_{1'}, \mathbf{e}_{2'}, \mathbf{e}_{3'}$, the form φ becomes

$$\varphi = a_{i'j'}x_{i'}y_{j'},$$

where

$$a_{i'j'} = \varphi(\mathbf{e}_{i'}, \mathbf{e}_{j'}).$$

But

$$\mathbf{e}_{i'} = \gamma_{i'i}\mathbf{e}_i$$

in going over to the new basis. It follows from the basic properties of a bilinear form that

$$a_{i'j'} = \varphi(\gamma_{i'i}\mathbf{e}_i, \gamma_{j'j}\mathbf{e}_j) = \gamma_{i'i}\gamma_{j'j}\varphi(\mathbf{e}_i, \mathbf{e}_j) = \gamma_{i'i}\gamma_{j'j}a_{ij}.$$

Hence the coefficients of the bilinear form φ transform according to the law

$$a_{i'j'} = \gamma_{i'i}\gamma_{j'j}a_{ij}. \tag{1}$$

Note the deep similarity between (1) and the transformation law (2), p. 39, for a linear form.

Conversely we have the following

THEOREM. *If the elements a_{ij} of the matrix* (1) *transform according to the law* (2) *under a basis transformation in L_3, then A is the matrix associated with a bilinear form.*

Proof. Let $\mathbf{e}_1, \mathbf{e}_2, \mathbf{e}_3$ and $\mathbf{e}_{1'}, \mathbf{e}_{2'}, \mathbf{e}_{3'}$ be two orthonormal bases in L_3, and let $\mathbf{x}, \mathbf{y}$ be any two vectors in L_3. Then

$$\mathbf{x} = x_i\mathbf{e}_i = x_{i'}\mathbf{e}_{i'}, \qquad \mathbf{y} = y_j\mathbf{e}_j = y_{j'}\mathbf{e}_{j'}.$$

Consider the expression $\varphi = a_{ij}x_iy_j$. To prove that φ is really a bilinear form defined on L_3, we must show that it does not change under a change of basis, i.e., that its value depends only on the vectors $\mathbf{x}, \mathbf{y}$ and not on the choice of basis. Under a change of basis φ goes over into $\varphi' = a_{i'j'}x_{i'}y_{j'}$. Hence we need only prove that $\varphi' = \varphi$. By formula (1) above and formula (9), p. 26,

$$\varphi' = a_{i'j'}x_{i'}y_{j'} = \gamma_{i'i}\gamma_{j'j}a_{ij}\gamma_{i'k}x_k\gamma_{j'l}y_l = \gamma_{i'i}\gamma_{i'k}\gamma_{j'j}\gamma_{j'l}a_{ij}x_ky_l.$$

Moreover,

$$\gamma_{i'i}\gamma_{i'k} = \delta_{ik}, \qquad \gamma_{j'j}\gamma_{j'l} = \delta_{jl},$$

by the properties of an orthogonal matrix,† and hence

$$\varphi' = \delta_{ik}\delta_{jl}a_{ij}x_ky_l.$$

But

$$\delta_{ik}x_k = x_i, \qquad \delta_{jl}y_l = y_j,$$

which implies

$$\varphi' = a_{ij}x_iy_j = \varphi. \quad ∎$$

† See formula (7), p. 23.

PROBLEMS

1. Prove that the coefficients of a bilinear form in the plane L_2 can be written as a square matrix of order two.

2. Write the scalar triple product $(\mathbf{a}, \mathbf{x}, \mathbf{y})$ figuring in Example 2 as a third-order determinant, and use the result to give another derivation of the coefficients of the corresponding bilinear form.

3. Let

$$\varphi[f(x), g(y)] = \int_a^b \int_a^b K(x, y) f(x) g(y)\, dx\, dy,$$

where $K(x, y)$ is a fixed function continuous in x and y. Is φ a bilinear form defined on the space $C[a, b]$ of all functions continuous in the interval $[a, b]$?

4. Let

$$\varphi[f(x), g(y)] = f(x_0) g(y_0),$$

where $a < x_0 < b$, $a < y_0 < b$. Is φ a bilinear form on the space $C[a, b]$?

5. Let x_1 and y_1 be the first components of the vectors $\mathbf{x}$ and $\mathbf{y}$ relative to some basis in the space L_n. Is the function

$$\varphi(\mathbf{x}, \mathbf{y}) = x_1^2 y_1$$

a bilinear form?

6. Is the function $\varphi(\mathbf{x}, \mathbf{y}) = c$ (c a fixed real number) a bilinear form?

10. Multilinear Forms. General Definition of a Tensor

10.1. A scalar function $\varphi = \varphi(\mathbf{x}, \mathbf{y}, \mathbf{z}, \ldots, \mathbf{w})$ of p vector arguments $\mathbf{x}, \mathbf{y}, \mathbf{z}, \ldots, \mathbf{w}$ is called a *multilinear function* or *multilinear form* if it is linear in all its arguments, i.e., if two conditions of the form

1) $\varphi(\mathbf{x}, \mathbf{y}, \mathbf{z}_1 + \mathbf{z}_2, \ldots, \mathbf{w}) = \varphi(\mathbf{x}, \mathbf{y}, \mathbf{z}_1, \ldots, \mathbf{w}) + \varphi(\mathbf{x}, \mathbf{y}, \mathbf{z}_2, \ldots, \mathbf{w})$,
2) $\varphi(\mathbf{x}, \mathbf{y}, \lambda\mathbf{z}, \ldots, \mathbf{w}) = \lambda\varphi(\mathbf{x}, \mathbf{y}, \mathbf{z}, \ldots, \mathbf{w})$

hold for each of the arguments $\mathbf{x}, \mathbf{y}, \mathbf{z}, \ldots, \mathbf{w}$. The number of arguments p is called the *degree* of the multilinear form, and φ itself is often called a *p-linear form*.

The linear forms considered in Sec. 8 are a special case of multilinear forms, i.e., forms of the first degree or 1-linear forms. Similarly, the bilinear forms considered in Sec. 9 are also a special case of multilinear forms, i.e., forms of the second degree or 2-linear forms. We now give some examples of multilinear forms of degree higher than two.

Example 1. The scalar triple product $(\mathbf{x}, \mathbf{y}, \mathbf{z})$ of three vectors $\mathbf{x}, \mathbf{y}, \mathbf{z} \in L_3$ is a trilinear (or 3-linear) form, since conditions of the type 1) and 2) hold for all three arguments $\mathbf{x}, \mathbf{y}, \mathbf{z}$.

Example 2. The product of three linear forms $\alpha(\mathbf{x})$, $\beta(\mathbf{y})$ and $\gamma(\mathbf{z})$ is a trilinear form. In fact, if

$$\varphi(\mathbf{x}, \mathbf{y}, \mathbf{z}) = \alpha(\mathbf{x})\beta(\mathbf{y})\gamma(\mathbf{z}),$$

then

$$\begin{aligned}\varphi(\mathbf{x}_1 + \mathbf{x}_2, \mathbf{y}, \mathbf{z}) &= \alpha(\mathbf{x}_1 + \mathbf{x}_2)\beta(\mathbf{y})\gamma(\mathbf{z}) = [\alpha(\mathbf{x}_1) + \alpha(\mathbf{x}_2)]\beta(\mathbf{y})\gamma(\mathbf{z}) \\ &= \alpha(\mathbf{x}_1)\beta(\mathbf{y})\gamma(\mathbf{z}) + \alpha(\mathbf{x}_2)\beta(\mathbf{y})\gamma(\mathbf{z}) = \varphi(\mathbf{x}_1, \mathbf{y}, \mathbf{z}) + \varphi(\mathbf{x}_2, \mathbf{y}, \mathbf{z}),\end{aligned}$$

$$\varphi(\lambda\mathbf{x}, \mathbf{y}, \mathbf{z}) = \alpha(\lambda\mathbf{x})\beta(\mathbf{y})\gamma(\mathbf{z}) = \lambda\alpha(\mathbf{x})\beta(\mathbf{y})\gamma(\mathbf{z}) = \lambda\varphi(\mathbf{x}, \mathbf{y}, \mathbf{z}),$$

and similarly for the other two arguments.

10.2. Next, given an orthonormal basis $\mathbf{e}_1, \mathbf{e}_2, \mathbf{e}_3$ in L_3, we find an expression for a p-linear form $\varphi(\mathbf{x}, \mathbf{y}, \mathbf{z}, \ldots, \mathbf{w})$ in terms of the components of $\mathbf{x}, \mathbf{y}, \mathbf{z}, \ldots, \mathbf{w}$ with respect to $\mathbf{e}_1, \mathbf{e}_2, \mathbf{e}_3$. For simplicity, we confine ourselves to the case of a trilinear form $\varphi(\mathbf{x}, \mathbf{y}, \mathbf{z})$. Let

$$\mathbf{x} = x_i\mathbf{e}_i, \qquad \mathbf{y} = y_j\mathbf{e}_j, \qquad \mathbf{z} = z_k\mathbf{e}_k,$$

where, as usual, we choose different indices merely for the convenience of subsequent calculations. By the linearity of φ in all three of its arguments,

$$\varphi(\mathbf{x}, \mathbf{y}, \mathbf{z}) = \varphi(x_i\mathbf{e}_i, y_j\mathbf{e}_j, z_k\mathbf{e}_k) = x_iy_jz_k\varphi(\mathbf{e}_i, \mathbf{e}_j, \mathbf{e}_k),$$

so that, writing

$$a_{ijk} = \varphi(\mathbf{e}_i, \mathbf{e}_j, \mathbf{e}_k),$$

we have

$$\varphi(\mathbf{x}, \mathbf{y}, \mathbf{z}) = a_{ijk}x_iy_jz_k.$$

This expression is a homogeneous polynomial of degree three, linear in all three sets of variables

$$x_1, x_2, x_3, \qquad y_1, y_2, y_3, \qquad z_1, z_2, z_3.$$

The polynomial contains $3^3 = 27$ terms and the same number of coefficients a_{ijk}. The coefficients a_{ijk} can be imagined as making up a "cubic array of order three."

In just the same way, a 4-linear form $\varphi(\mathbf{x}, \mathbf{y}, \mathbf{z}, \mathbf{u})$ can be written as

$$\varphi = a_{ijkl}x_iy_jz_ku_l$$

(in the basis $\mathbf{e}_1, \mathbf{e}_2, \mathbf{e}_3$), where

$$a_{ijkl} = \varphi(\mathbf{e}_i, \mathbf{e}_j, \mathbf{e}_k, \mathbf{e}_l)$$

and the corresponding polynomial has 3^4 terms and the same number of coefficients a_{ijkl}. More generally, a p-linear form $\varphi(\mathbf{x}, \mathbf{y}, \mathbf{z}, \ldots, \mathbf{w})$ can be written as

$$\varphi = a_{ijk\cdots m}x_iy_jz_k \cdots w_m,$$

where

$$a_{ijk\cdots m} = \varphi(\mathbf{e}_i, \mathbf{e}_j, \mathbf{e}_k, \ldots, \mathbf{e}_m). \tag{1}$$

The coefficients $a_{ijk\cdots m}$ of this form have p indices, each of which can take three values 1, 2, 3. Hence a p-linear form has 3^p coefficients in all.

Example 1'. The trilinear form $(\mathbf{x}, \mathbf{y}, \mathbf{z})$ considered in Example 1 becomes

$$(\mathbf{x}, \mathbf{y}, \mathbf{z}) = \epsilon_{ijk}x_i y_j z_k$$

in component form (see p. 19), i.e., the general coefficient is just the antisymmetric Kronecker symbol introduced on p. 17.

Example 2'. In the case of the trilinear form considered in Example 2, suppose the linear forms $\alpha(\mathbf{x})$, $\beta(\mathbf{y})$, $\gamma(\mathbf{z})$ are

$$\alpha(\mathbf{x}) = a_i x_i, \qquad \beta(\mathbf{y}) = b_j y_j, \qquad \gamma(\mathbf{z}) = c_k z_k$$

in the orthonormal basis $\mathbf{e}_1, \mathbf{e}_2, \mathbf{e}_3$. Then the trilinear form $\varphi(\mathbf{x}, \mathbf{y}, \mathbf{z}) = \alpha(\mathbf{x})\beta(\mathbf{y})\gamma(\mathbf{z})$ becomes

$$\varphi(\mathbf{x}, \mathbf{y}, \mathbf{z}) = a_i b_j c_k x_i y_j z_k,$$

with coefficients

$$a_{ijk} = a_i b_j c_k.$$

10.3. The definition of a multilinear form $\varphi = \varphi(\mathbf{x}, \mathbf{y}, \mathbf{z}, \ldots, \mathbf{w}, \ldots)$ is independent of the choice of coordinate system, i.e., the value of φ depends only on the values of its vector arguments. For example, a trilinear form $\varphi = \varphi(\mathbf{x}, \mathbf{y}, \mathbf{z})$ depends only on the values of the vectors $\mathbf{x}, \mathbf{y}, \mathbf{z}$ and not on the components of $\mathbf{x}, \mathbf{y}, \mathbf{z}$ relative to any underlying basis $\mathbf{e}_1, \mathbf{e}_2, \mathbf{e}_3$. In the language of p. 25, we can say that multilinear forms have been defined in an *invariant* fashion.

Since the components of a vector change in transforming to a new basis, the same must be true of the coefficients of a multilinear form (if the form itself is to remain invariant). The set of coefficients of an invariant multilinear form constitutes a very important geometrical object:

DEFINITION. *The geometric (or physical) object specified by the set of coefficients $a_{ijk\cdots m}$ of a multilinear form $\varphi = \varphi(\mathbf{x}, \mathbf{y}, \mathbf{z}, \ldots, \mathbf{w})$ written in some orthonormal basis is called an **orthogonal tensor**, and the numbers $a_{ijk\cdots m}$ themselves are called the **components** of the tensor.*

Remark 1. The tensors considered in this book are all orthogonal, and hence the term "tensor" will always refer to an "orthogonal tensor."

Remark 2. The tensor $a_{ijk\cdots m}$ is said to be *determined* by the multilinear form $\varphi = \varphi(\mathbf{x}, \mathbf{y}, \mathbf{z}, \ldots, \mathbf{w})$. The coefficients $a_{ijk\cdots m}$ of a form φ of degree p are given by formula (1) and have p indices. Correspondingly, a tensor determined by a form of degree p is called a tensor of *order p*.

Example 1. If φ is a trilinear form defined on L_3, then each index of the corresponding tensor can independently take the values 1, 2 and 3. Hence a tensor of order p in three-dimensional space has 3^p components. By the same token, such a tensor has 2^p components in the plane and n^p components in the n-dimensional space L_n.

Example 2. The coefficients a_i of a linear form $\varphi = \varphi(\mathbf{x})$ constitute a first-order tensor. Moreover, since the scalar product of an arbitrary constant vector $\mathbf{a}$ with a variable vector $\mathbf{x}$ is a linear form, the components a_i of any vector $\mathbf{a}$ also constitute a first-order tensor.

Example 3. In just the same way, the coefficients a_{ij} of a bilinear form $\varphi = \varphi(\mathbf{x}, \mathbf{y})$, making up a matrix $A = (a_{ij})$, constitute a second-order tensor. In particular, since

$$\mathbf{x} \cdot \mathbf{y} = \delta_{ij} x_i y_j,$$

where $\mathbf{x} \cdot \mathbf{y}$ is the scalar product of two vectors $\mathbf{x}$ and $\mathbf{y}$ with components x_i and y_j in some orthonormal basis, the values of the symmetric Kronecker symbol δ_{ij} are the coefficients of a bilinear form. Hence δ_{ij} is a second-order tensor, known as the *unit tensor*.

Example 4. Since

$$(\mathbf{x}, \mathbf{y}, \mathbf{z}) = \epsilon_{ijk} x_i y_j z_k,$$

where x_i, y_j and z_k are the components of the vectors $\mathbf{x}$, $\mathbf{y}$ and $\mathbf{z}$ in some orthonormal basis, the values of the antisymmetric Kronecker symbol ϵ_{ijk} are the coefficients of a trilinear form. Hence ϵ_{ijk} is a third-order tensor, known as the *discriminantal tensor*.

Example 5. A scalar quantity, i.e., a quantity independent of the choice of the underlying basis, is called a *tensor of order zero* and can be thought of as the unique coefficient of a linear form of degree zero. A tensor of order zero is also called an *invariant*, since its unique component does not change under basis transformations.

Two tensors are said to be *equal* if the multilinear forms determining then are identical. Equal tensors have the same order, and their components are equal in any coordinate system. In fact, the identity

$$\varphi(\mathbf{x}, \mathbf{y}, \mathbf{z}, \ldots, \mathbf{w}) = \psi(\mathbf{x}, \mathbf{y}, \mathbf{z}, \ldots, \mathbf{w})$$

becomes

$$a_{ijk\cdots m} x_i y_j z_k \cdots w_m = b_{ijk\cdots m} x_i y_j z_k \cdots w_m$$

in component form, which immediately implies

$$a_{ijk\cdots m} = b_{ijk\cdots m}.$$

If the multilinear form $\varphi = \varphi(\mathbf{x}, \mathbf{y}, \mathbf{z}, \ldots, \mathbf{w})$ is identically zero, then the tensor determined by φ is called the *null tensor*. The components of the null tensor are clearly all zero.

10.4. In going over to a new basis, the components of the vectors making up the arguments of a multilinear form transform in the way described by formula (9), p. 26. Hence the coefficients of the form, i.e., the components of the tensor determined by the form, must also transform in some perfectly well-defined way. This transformation law is given by the following

THEOREM. *A set of quantities* $a_{ijk\cdots m}$ *depending on the choice of basis forms a tensor if and only if they transform according to the law*

$$a_{i'j'k'\cdots m'} = \gamma_{i'i}\gamma_{j'j}\gamma_{k'k}\cdots\gamma_{m'm}a_{ijk\cdots m} \tag{2}$$

under the transformation from one orthonormal basis $\mathbf{e}_1, \mathbf{e}_2, \mathbf{e}_3$ *to another orthonormal basis* $\mathbf{e}_{1'}, \mathbf{e}_{2'}, \mathbf{e}_{3'}$.

Proof. Suppose $a_{ijk\cdots m}$ is a tensor. Then the quantities $a_{ijk\cdots m}$ are the coefficients of some multilinear form $\varphi = \varphi(\mathbf{x}, \mathbf{y}, \mathbf{z}, \ldots, \mathbf{w})$, and hence

$$a_{ijk\cdots m} = \varphi(\mathbf{e}_i, \mathbf{e}_j, \mathbf{e}_k, \ldots, \mathbf{e}_m).$$

The coefficients of φ in the new basis are given by the analogous formula

$$a_{i'j'k'\cdots m'} = \varphi(\mathbf{e}_{i'}, \mathbf{e}_{j'}, \mathbf{e}_{k'}, \ldots, \mathbf{e}_{m'}).$$

But the vectors $\mathbf{e}_{i'}$ of the new basis are expressed in terms of the vectors $\mathbf{e}_i$ of the old basis by formula (1), p. 22,

$$\mathbf{e}_{i'} = \gamma_{i'i}\mathbf{e}_i,$$

and hence

$$a_{i'j'k'\cdots m'} = \varphi(\gamma_{i'i}\mathbf{e}_i, \gamma_{j'j}\mathbf{e}_j, \gamma_{k'k}\mathbf{e}_k, \ldots, \gamma_{m'm}\mathbf{e}_m).$$

Since the form φ is multilinear, it follows that

$$\begin{aligned} a_{i'j'k'\cdots m'} &= \gamma_{i'i}\gamma_{j'j}\gamma_{k'k}\cdots\gamma_{m'm}\varphi(\mathbf{e}_i, \mathbf{e}_j, \mathbf{e}_k, \ldots, \mathbf{e}_m) \\ &= \gamma_{i'i}\gamma_{j'j}\gamma_{k'k}\cdots\gamma_{m'm}a_{ijk\cdots m}, \end{aligned}$$

in keeping with (2).

Conversely, suppose the quantities $a_{ijk\cdots m}$ transform in accordance with (2) in going over to a new basis. Suppose $a_{ijk\cdots m}$ has p indices, and let $\mathbf{x}, \mathbf{y}, \mathbf{z}, \ldots, \mathbf{w}$ be p vectors whose expansions relative to the old and new bases are given by

$$\mathbf{x} = x_i\mathbf{e}_i = x_{i'}\mathbf{e}_{i'}, \qquad \mathbf{y} = y_j\mathbf{e}_j = y_{j'}\mathbf{e}_{j'},$$

$$\mathbf{z} = z_k\mathbf{e}_k = z_{k'}\mathbf{e}_{k'}, \ldots, \mathbf{w} = w_m\mathbf{e}_m = w_{m'}\mathbf{e}_{m'}.$$

To prove that the set of quantities $a_{ijk\cdots m}$ forms a tensor, we must show that the expression

$$\varphi = a_{ijk\cdots m}x_iy_jz_k\cdots w_m \tag{3}$$

is a multilinear form, i.e., that it depends only on the choice of the vectors $\mathbf{x}, \mathbf{y}, \mathbf{z}, \ldots, \mathbf{w}$ and not on the choice of basis. But (3) becomes

$$\varphi' = a_{i'j'k'\cdots m'}x_{i'}y_{j'}z_{k'}\cdots w_{m'} \tag{3'}$$

under the basis transformation. Substituting (2) into (3') and replacing

$x_{i'}, y_{j'}, z_{k'}, \ldots, w_{m'}$ by the values given by formula (9), p. 26 and its analogues, we get

$$\begin{aligned}\varphi' &= \gamma_{i'i}\gamma_{j'j}\gamma_{k'k} \cdots \gamma_{m'm}a_{ijk\cdots m}\gamma_{i'p}x_p\gamma_{j'q}y_q\gamma_{k'r}z_r \cdots \gamma_{m's}w_s \\ &= (\gamma_{i'i}\gamma_{i'p})(\gamma_{j'j}\gamma_{j'q})(\gamma_{k'k}\gamma_{k'r}) \cdots (\gamma_{m'm}\gamma_{m's})a_{ijk\cdots m}x_py_qz_r \cdots w_s.\end{aligned}$$

But by the orthogonality relations (7), p. 23,

$$\gamma_{i'i}\gamma_{i'p} = \delta_{ip}, \quad \gamma_{j'j}\gamma_{j'q} = \delta_{jq}, \quad \gamma_{k'k}\gamma_{k'r} = \delta_{kr}, \ldots, \quad \gamma_{m'm}\gamma_{m's} = \delta_{ms},$$

and hence

$$\varphi' = a_{ijk\cdots m}\delta_{ip}x_p\delta_{jq}y_q\delta_{kr}z_r \cdots \delta_{ms}w_s = a_{ijk\cdots m}x_iy_jz_k \cdots w_m = \varphi. \quad ■$$

PROBLEMS

1. Is the function $\varphi(x, y, z, \ldots, w) = c$ (c a fixed real number) a multilinear form?

2. Let x_1, y_1 and z_1 be the first components of the vectors x, y and z relative to some basis in the space L_n. Is the function

$$\varphi(x, y, z) = x_1^2 y_1 z_1$$

a trilinear form?

3. Let

$$\varphi[f_1(t), f_2(t), \ldots, f_k(t)] = f_1(t_1)f_2(t_2) \cdots f_k(t_k),$$

where $a < t_i < b$ ($i = 1, \ldots, k$). Is φ a multilinear form defined on the space $C[a, b]$ of all functions continuous in the interval $[a, b]$?

4. Let

$$\varphi[f(x), g(y), h(z)] = \int_a^b \int_a^b \int_a^b K(x, y, z)f(x)g(y)h(z)\, dx\, dy\, dz,$$

where $K(x, y, z)$ is a fixed function continuous in x, y and z. Is φ a trilinear form on $C[a, b]$?

5. Suppose $\mathbf{x} = x_i\mathbf{e}_i$ relative to some orthonormal basis $\mathbf{e}_1, \mathbf{e}_2, \mathbf{e}_3$ in L_3. Prove that the numbers $x_{ij} = x_ix_j$ form a second-order tensor.

6. Prove that the components of the unit tensor δ_{ij} have the same values in all orthonormal bases, i.e., that $\delta_{i'j'} = \delta_{ij}$ if $i' = i, j' = j$.

7. Prove that the components of the discriminantal tensor ϵ_{ijk} have the same values in all orthonormal bases with the same orientation and the negative of these values in bases with the opposite orientation, i.e., that $\epsilon_{i'j'k'} = \pm\epsilon_{ijk}$ if $i' = i, j' = j, k' = k$.

8. Prove that the set of quantities α_{ijkl}, defined in every orthonormal basis $\mathbf{e}_1, \mathbf{e}_2, \mathbf{e}_3$ as

$$\alpha_{ijkl} = \begin{cases} 1 \text{ if } i = k, j = l, \\ \text{otherwise,} \end{cases}$$

forms a tensor of order 4.

9. Write the transformation law for the components of a tensor of order 5.

10. Let $\varphi(x_1, \ldots, x_n)$ be an invariant function of the rectangular coordinates x_i. Prove that the quantities

$$\frac{\partial \varphi}{\partial x_i}$$

form a first-order tensor, while the quantities

$$\frac{\partial^2 \varphi}{\partial x_i \partial x_j}$$

form a second-order tensor.

11. Algebraic Operations on Tensors

11.1. Addition of tensors. Let $\varphi = \varphi(\mathbf{x}, \mathbf{y}, \mathbf{z}, \ldots, \mathbf{w})$ and $\psi = \psi(\mathbf{x}, \mathbf{y}, \mathbf{z}, \ldots, \mathbf{w})$ be two multilinear forms of the same degree p in the same vector arguments. Then the sum $\varphi + \psi$ is clearly a multilinear form of the same degree. By the *sum* of the tensors $a_{ijk\cdots m}$ and $b_{ijk\cdots m}$ of order p determined by the forms φ and ψ we mean the tensor $c_{ijk\cdots m}$ determined by the form $\varphi + \psi$. Since

$$\varphi + \psi = (a_{ijk\cdots m} + b_{ijk\cdots m})x_i y_j \cdots w_m,$$

the components of the tensor $c_{ijk\cdots m}$ are connected with those of the tensors $a_{ijk\cdots m}$ and $b_{ijk\cdots m}$ by the relation

$$c_{ijk\cdots m} = a_{ijk\cdots m} + b_{ijk\cdots m}.$$

11.2. Multiplication of tensors by real numbers. The product $\lambda\varphi$ of a real number λ and a multilinear form φ of degree p is again a multilinear form of degree p. By the *product* of λ and the tensor $a_{ijk\cdots m}$ of order p determined by the form φ we mean the tensor $b_{ijk\cdots m}$ of the same order determined by the form $\lambda\varphi$. Since

$$\lambda\varphi = (\lambda a_{ijk\cdots m})x_i y_j z_k \cdots w_m,$$

we have

$$b_{ijk\cdots m} = \lambda a_{ijk\cdots m}.$$

Remark. It follows from the foregoing that the set of all multilinear forms of degree p, as well as the set of all tensors of order p, forms a linear space. The dimension of this space is just 3^p, with a basis consisting, say, of the 3^p p-linear forms

$$\varphi_{ijk\cdots m} = x_i y_j z_k \cdots w_m.$$

11.3. Multiplication of tensors. Let φ and ψ be two multilinear forms of degrees p and q, respectively, with different vector arguments. Then the product $\varphi\psi$ is clearly a multilinear form of degree $p + q$. For example, if

$\varphi = \varphi(\mathbf{x}, \mathbf{y}, \mathbf{z})$ is a trilinear form and $\psi = \psi(\mathbf{u}, \mathbf{v})$ a bilinear form, then the product $\varphi\psi = \varphi(\mathbf{x}, \mathbf{y}, \mathbf{z})\psi(\mathbf{u}, \mathbf{v})$ is a multilinear form of degree 5.

The forms φ and ψ determine tensors of orders p and q, respectively. By the *product* of the tensors determined by the forms φ and ψ we mean the tensor determined by the product $\varphi\psi$. Since the form $\varphi\psi$ is of degree $p + q$, the product of two tensors of orders p and q is a tensor of order $p + q$. For example, the forms

$$\varphi(\mathbf{x}, \mathbf{y}, \mathbf{z}) = a_{ijk}x_iy_jz_k$$

and

$$\psi(\mathbf{u}, \mathbf{v}) = b_{lm}u_lv_m$$

determine tensors a_{ijk} and b_{lm} of orders 3 and 2, respectively, and their product

$$\varphi(\mathbf{x}, \mathbf{y}, \mathbf{z})\psi(\mathbf{u}, \mathbf{v}) = (a_{ijk}b_{lm})x_iy_jz_ku_lv_m$$

determines a tensor $a_{ijk}b_{lm}$ of order 5, i.e., the product of the tensors a_{ijk} and b_{lm}.

Remark. In Example 3′, p. 42, we in effect constructed the second-order tensor equal to the product of two first-order tensors a_i and b_j. Similarly, in Example 2′, p. 46, we constructed the third-order tensor equal to the product of three first-order tensors a_i, b_j and c_k.

11.4. Contraction of tensors. Given a multilinear form $\varphi = \varphi(\mathbf{x}, \mathbf{y}, \mathbf{z}, \ldots, \mathbf{w})$ of degree p, suppose we replace any two arguments, say $\mathbf{x}$ and $\mathbf{y}$, by the basis vectors $\mathbf{e}_i$ and $\mathbf{e}_j$, writing

$$\varphi(\mathbf{e}_i, \mathbf{e}_j, \mathbf{z}, \ldots, \mathbf{w}) = \varphi_{ij}.$$

Then φ_{ij} is a linear function of the vector arguments $\mathbf{z}, \ldots, \mathbf{w}$, but not a linear form, since it now depends on the choice of basis. To determine how φ_{ij} changes under basis transformations in L_3, let

$$\varphi_{i'j'} = \varphi(\mathbf{e}_{i'}, \mathbf{e}_{j'}, \mathbf{z}, \ldots, \mathbf{w}).$$

Then, since

$$\mathbf{e}_{i'} = \gamma_{i'i}\mathbf{e}_i, \qquad \mathbf{e}_{j'} = \gamma_{j'j}\mathbf{e}_{j'},$$

we have

$$\begin{aligned}\varphi_{i'j'} = \varphi(\gamma_{i'i}\mathbf{e}_i, \gamma_{j'j}\mathbf{e}_j, \mathbf{z}, \ldots, \mathbf{w}) &= \gamma_{i'i}\gamma_{j'j}\varphi(\mathbf{e}_i, \mathbf{e}_j, \mathbf{z}, \ldots, \mathbf{w}) \\ &= \gamma_{i'i}\gamma_{j'j}\varphi_{ij}.\end{aligned}$$

Suppose we set $i' = j'$ and then sum over the resulting expressions. This gives

$$\varphi_{i'i'} = \gamma_{i'i}\gamma_{i'j}\varphi_{ij}.$$

But

$$\gamma_{i'i}\gamma_{i'j} = \delta_{ij}$$

by the orthogonality relations (7), p. 23, and hence

$$\varphi_{i'i'} = \delta_{ij}\varphi_{ij} = \varphi_{ii}.$$

It follows that the expression φ_{ii}, which is linear in the vector arguments $\mathbf{z}, \ldots, \mathbf{w}$, does not depend on the choice of basis. Hence φ_{ii} is a multilinear form in $\mathbf{z}, \ldots, \mathbf{w}$, in fact a form of degree $p - 2$, since it depends on two fewer vector arguments than the original form φ.

Writing φ in component form, we get

$$\varphi = \varphi(\mathbf{x}, \mathbf{y}, \mathbf{z}, \ldots, \mathbf{w}) = a_{ijk\cdots m}x_i y_j z_k \cdots w_m.$$

The substitution $\mathbf{x} = \mathbf{e}_i$, $\mathbf{y} = \mathbf{e}_j$ then gives

$$\varphi_{ij} = \varphi(\mathbf{e}_i, \mathbf{e}_j, \mathbf{z}, \ldots, \mathbf{w}) = a_{ijk\cdots m}z_k \ldots w_m,$$

since in this case

$$x_i = 1, \quad x_p = 0 \quad \text{if} \quad p \neq i,$$

$$y_j = 1, \quad y_q = 0 \quad \text{if} \quad q \neq j.$$

It follows that

$$\varphi_{ii} = a_{iik\cdots m}z_k \cdots w_m.$$

Hence the expression for the components of the tensor $b_{k\cdots m}$ of order $p - 2$ determined by the form φ_{ii} in terms of the components of the tensor $a_{ijk\cdots m}$ determined by the original form φ is just

$$b_{k\cdots m} = a_{iik\cdots m},$$

or, in more detail,

$$b_{k\cdots m} = a_{11k\cdots m} + a_{22k\cdots m} + a_{33k\cdots m}.$$

The operation leading from the tensor $a_{ijk\cdots m}$ to the tensor $b_{k\cdots m}$ is called *contraction* of $a_{ijk\cdots m}$ with respect to the indices i and j.

In just the same way, we define contraction of the tensor $a_{ijk\cdots m}$ with respect to any other pair of indices. As just shown, *contraction of a tensor lowers its order by two.* For example, contraction of a second-order tensor a_{ij} leads to a tensor a_{ii} of order zero, i.e., to an *invariant*. This invariant is called the *trace* of the tensor a_{ij}, denoted by

$$a_{ii} = \operatorname{tr}(a_{ij}).$$

11.5. Contraction of products of tensors. Given a product of two tensors, e.g., the tensors a_{ijk} and b_{lm} (of orders 3 and 2, respectively), suppose we form the product $a_{ijk}b_{lm}$ (a tensor of order 5), and then contract the resulting tensor with respect to the indices k and l, say. This gives a tensor

$$a_{ijk}b_{km} = a_{ij1}b_{1m} + a_{ij2}b_{2m} + a_{ij3}b_{3m}$$

of order 3, and the corresponding operation is again called *contraction*, more exactly, contraction of the tensors a_{ijk} and b_{lm} with respect to the indices k and l. Thus the operation of contracting two tensors consists of first multiplying them and then contracting the resulting tensor with respect to a pair of indices, one belonging to each factor. *Contraction of two tensors, one of order p and the other of order q, clearly gives a tensor of order* $p + q - 2$.

Remark 1. In effect, the operation of contraction of tensors has already been encountered many times. For example, the scalar product of two vectors $\mathbf{x} = x_i\mathbf{e}_i$ and $\mathbf{y} = y_j\mathbf{e}_j$, given by the formula

$$\mathbf{x}\cdot\mathbf{y} = x_i y_i,$$

is just the result of contracting the two first-order tensors x_i and y_i formed from the components of the vectors $\mathbf{x}$ and $\mathbf{y}$. The linear form

$$\varphi(\mathbf{x}) = a_i x_i$$

is the result of contracting the tensors a_i and x_i, the bilinear form

$$\varphi(\mathbf{x}, \mathbf{y}) = a_{ij}x_i x_j$$

is the result of first contracting the tensor a_{ij} with the tensor x_i and then contracting the tensor $a_{ij}x_i$ with the tensor y_j, and so on. More generally, as the last example makes clear, we can contract a product of tensors not only with respect to one pair of indices, but also with respect to any r pairs of indices. The result is a new tensor whose order is $2r$ less than the sum of the orders of the original tensors.

Remark 2. A particularly simple result is obtained if we contract an arbitrary tensor with the unit tensor. For example,

$$a_{ijk}\delta_{kl} = a_{ij1}\delta_{1l} + a_{ij2}\delta_{2l} + a_{ij3}\delta_{3l} = a_{ijl},$$

since

$$\delta_{kl} = \begin{cases} 1 \text{ if } k = l, \\ 0 \text{ if } k \neq l. \end{cases}$$

We now prove an important *indirect test for tensor character*:

THEOREM. *Let*

$$a_{i_1\cdots i_p j_1\cdots j_q} \tag{1}$$

be a set of 3^{p+q} *numbers specified in every orthonormal basis in* L_3, *and suppose contraction of* (1) *with an arbitrary tensor* $t_{j_1\cdots j_q}$ *of order q gives another tensor of order p. Then* (1) *is a tensor of order* $p + q$.

Proof. For simplicity, consider the special case $p = 3$, $q = 2$, where the set of numbers (1) is of the form a_{ijklm}. Suppose the quantity

$$s_{ijk} = a_{ijklm}t_{lm}$$

is a tensor whenever t_{lm} is a tensor. Let $t_{lm} = u_l v_m$ (the product of the vectors u_l and v_m). Then

$$s_{ijk} = a_{ijklm}u_l v_m,$$

and contracting this expression with arbitrary vectors x_i, y_j, z_k, we get

$$s_{ijk}x_i y_j z_k = a_{ijklm}x_i y_j z_k u_l v_m.$$

Since s_{ijk} is a tensor, the expression on the left is a scalar. It follows that the expression on the right, which depends linearly on the components of the vectors **x**, **y**, **z**, **u**, **v**, is a multilinear form of degree 5. But the numbers a_{ijklm} are the coefficients of this form, and hence make up a tensor of order 5. This proves the theorem for $p = 3, q = 2$. The proof is virtually the same for general p and q. ∎

11.6. Permutation of indices. Let $a_{ijk\cdots m}$ be the tensor determined by the multilinear form $\varphi = \varphi(\mathbf{x}, \mathbf{y}, \mathbf{z}, \ldots, \mathbf{w})$, so that

$$\varphi = a_{ijk\cdots m}x_iy_jz_k \cdots w_m,$$

and consider the form ψ obtained from φ by permuting some of its arguments. For example, suppose

$$\psi(\mathbf{x}, \mathbf{y}, \mathbf{z}, \ldots, \mathbf{w}) = \varphi(\mathbf{y}, \mathbf{z}, \mathbf{x}, \ldots, \mathbf{w}). \tag{2}$$

If $b_{ijk\cdots m}$ denotes the tensor determined by ψ, we can write (2) in the form

$$b_{ijk\cdots m}x_iy_jz_k \cdots w_m = a_{ijk\cdots m}y_iz_jx_k \cdots w_m. \tag{3}$$

Changing indices of summation in the right-hand side, and bearing in mind that (3) is an identity, we get

$$b_{ijk\cdots m} = a_{ijk\cdots m}.$$

The tensor $b_{ijk\cdots m}$ differs from the tensor $a_{ijk\cdots m}$ only in the arrangement of its indices. Thus permutation of the indices of a tensor leads to another tensor. It is important to note that the tensors $a_{ijk\cdots m}$ are actually distinct, since corresponding components of the two tensors (i.e., components with identical indices) are in general unequal.

PROBLEMS

1. Given a second-order tensor a_{ij}, prove that the cofactors A_{ij} of the determinant a made up of the components of a_{ij} is also a second-order tensor, satisfying the relation

$$A_{ik}a_{kj} = a\delta_{ij}$$

(cf. Sec. 5, Prob. 5).

2. Use multiplication and subsequent contraction to construct tensors of orders 5, 3 and 1 from a given third-order tensor a_{ijk} and second-order tensor b_{lm}.

3. Prove that the second-order tensor z_{ij} is a product of two first-order tensors if and only if its components satisfy the condition

$$z_{ij}z_{kl} - z_{il}z_{kj} = 0.$$

4. Construct an invariant by contraction of the tensor a_{ij} whose components are the elements of the matrix

$$\begin{pmatrix} 2 & 1 & 0 \\ 3 & -5 & 6 \\ -7 & 0 & 4 \end{pmatrix}.$$

5. Let a_{ij} be a second-order tensor with matrix

$$(a_{ij}) = \begin{pmatrix} 2 & 0 & 3 \\ 5 & 1 & 2 \\ 4 & 5 & 7 \end{pmatrix}$$

in some basis, and let x_i and y_j be first-order tensors (vectors) with components 2, 1, 4 and 3, 7, −1, respectively, in the same basis. Find

a) $a_{ij}x_j$; b) $a_{ij}x_i$; c) $a_{ij}y_i$; d) $a_{ij}y_j$; e) $a_{ij}x_iy_j$; f) $a_{ij}y_ix_j$;
g) $a_{ij}\delta_{ij}$; h) $a_{ij} - \frac{2}{3}\delta_{ij}a_{ll}$; i) $(a_{ij} - \frac{2}{3}\delta_{ij}a_{ll})x_i$; j) $(a_{ij} - \frac{2}{3}\delta_{ij}a_{ll})x_iy_j$.

6. Find a basis for the linear space consisting of all second-order tensors.

12. Symmetric and Antisymmetric Tensors

12.1. Let $\varphi = \varphi(\mathbf{x}, \mathbf{y})$ be a bilinear form. Then φ is said to be *symmetric* if

$$\varphi(\mathbf{x}, \mathbf{y}) = \varphi(\mathbf{y}, \mathbf{x})$$

for all **x** and **y**. A second-order tensor determined by a symmetric bilinear form is also said to be *symmetric*. The components of a symmetric second-order tensor in any orthonormal basis form a *symmetric matrix*, i.e., satisfy the condition

$$a_{ij} = a_{ji}. \tag{1}$$

Since

$$a_{ij} = \varphi(\mathbf{e}_i, \mathbf{e}_j),$$

(1) follows from the fact that

$$\varphi(\mathbf{e}_i, \mathbf{e}_j) = \varphi(\mathbf{e}_j, \mathbf{e}_i).$$

Conversely, if (1) holds, then the bilinear form $\varphi(\mathbf{x}, \mathbf{y}) = a_{ij}x_iy_j$ is symmetric since then

$$\varphi(\mathbf{x}, \mathbf{y}) = a_{ij}x_iy_j = a_{ji}x_iy_j = a_{ij}y_ix_j = \varphi(\mathbf{y}, \mathbf{x}).$$

Clearly, every symmetric matrix (a_{ij}) is of the form

$$\begin{pmatrix} a_{11} & a_{12} & a_{13} \\ a_{12} & a_{22} & a_{23} \\ a_{13} & a_{23} & a_{33} \end{pmatrix},$$

where there are only six distinct matrix elements and, by the same token, only six distinct components of the corresponding tensor.

Example 1. The scalar product of two vectors $\mathbf{x}$ and $\mathbf{y}$ is a symmetric bilinear form, since

$$\mathbf{x}\cdot\mathbf{y} = \mathbf{y}\cdot\mathbf{x}.$$

The coefficients of this form make up the unit tensor δ_{ij}, whose matrix

$$\begin{pmatrix} 1 & 0 & 0 \\ 0 & 1 & 0 \\ 0 & 0 & 1 \end{pmatrix}$$

is obviously symmetric.

More generally, let $\varphi = \varphi(\mathbf{x}, \mathbf{y}, \mathbf{z}, \ldots, \mathbf{w})$ be a multilinear form of degree p. Then φ is said to be *symmetric in two (given) arguments* if it does not change value when the two arguments are interchanged. By the same token, the tensor determined by φ is said to be *symmetric in the corresponding indices.* For example, we say that the form φ is symmetric in $\mathbf{x}$ and $\mathbf{z}$ if

$$\varphi(\mathbf{x}, \mathbf{y}, \mathbf{z}, \ldots, \mathbf{w}) = \varphi(\mathbf{z}, \mathbf{y}, \mathbf{x}, \ldots, \mathbf{w}),$$

and the tensor $a_{ijk\cdots m}$ determined by φ is symmetric in the indices i and k, so that its components satisfy the condition

$$a_{ijk\cdots m} = a_{kji\cdots m}$$

in every coordinate system.

A multilinear form of degree p is said to be *symmetric* if it does not change under any permutation of its arguments, and the corresponding tensor is called a *symmetric tensor* of order p. Thus arbitrarily rearranging the indices of a component of a symmetric tensor has no effect on its value.

Example 2. The trilinear form $\varphi(\mathbf{x}, \mathbf{y}, \mathbf{z})$ is symmetric if and only if

$$\varphi(\mathbf{x}, \mathbf{y}, \mathbf{z}) = \varphi(\mathbf{y}, \mathbf{z}, \mathbf{x}) = \varphi(\mathbf{z}, \mathbf{x}, \mathbf{y}) = \varphi(\mathbf{y}, \mathbf{x}, \mathbf{z}) = \varphi(\mathbf{z}, \mathbf{y}, \mathbf{x}) = \varphi(\mathbf{x}, \mathbf{z}, \mathbf{y})$$

for arbitrary vectors $\mathbf{x}$, $\mathbf{y}$ and $\mathbf{z}$. The components of the tensor a_{ijk} determined by φ do not change under arbitrary permutations of indices.

12.2. A bilinear form $\varphi = \varphi(\mathbf{x}, \mathbf{y})$ is said to be *antisymmetric* if

$$\varphi(\mathbf{x}, \mathbf{y}) = -\varphi(\mathbf{y}, \mathbf{x})$$

for all $\mathbf{x}$ and $\mathbf{y}$. A second-order tensor determined by an antisymmetric form is also said to be *antisymmetric*. Since $\varphi(\mathbf{e}_i, \mathbf{e}_j) = -\varphi(\mathbf{e}_j, \mathbf{e}_i)$, the components of an antisymmetric second-order tensor satisfy the condition

$$a_{ij} = -a_{ji}$$

in any basis, i.e., form a *skew-symmetric matrix* of the form

$$(a_{ij}) = \begin{pmatrix} 0 & a_{12} & -a_{31} \\ -a_{12} & 0 & a_{23} \\ a_{31} & -a_{23} & 0 \end{pmatrix}. \tag{2}$$

It is apparent from (2) that an antisymmetric second-order tensor has in effect only three components.

More generally, let $\varphi = \varphi(\mathbf{x}, \mathbf{y}, \mathbf{z}, \ldots, \mathbf{w})$ be a multilinear form of degree. Then φ is said to be *antisymmetric in two (given) arguments* if it changes sign when the two arguments are interchanged. By the same token, the tensor determined by φ is *antisymmetric in the corresponding indices*. A multilinear form of degree p is said to be *antisymmetric* (without further qualification) if it changes sign when any two of its arguments are interchanged, and the tensor determined by such a form is called an *antisymmetric tensor* of order p. Thus an antisymmetric tensor changes sign when any two of its indices are interchanged.

Example. The scalar triple product $(\mathbf{x}, \mathbf{y}, \mathbf{z})$ of three vectors $\mathbf{x}$, $\mathbf{y}$ and $\mathbf{z}$ is an antisymmetric trilinear form, and the tensor ϵ_{ijk} determined by this form (the discriminantal tensor) is an antisymmetric tensor with effectively only one component $\epsilon_{123} = \epsilon$.

12.3. Suppose we use a given bilinear form $\varphi = \varphi(\mathbf{x}, \mathbf{y})$ to construct the two related bilinear forms

$$\varphi_1(\mathbf{x}, \mathbf{y}) = \tfrac{1}{2}[\varphi(\mathbf{x}, \mathbf{y}) + \varphi(\mathbf{y}, \mathbf{x})],$$
$$\varphi_2(\mathbf{x}, \mathbf{y}) = \tfrac{1}{2}[\varphi(\mathbf{x}, \mathbf{y}) - \varphi(\mathbf{y}, \mathbf{x})].$$

Then φ_1 is clearly symmetric, while φ_2 is antisymmetric. In fact,

$$\varphi_1(\mathbf{x}, \mathbf{y}) = \tfrac{1}{2}[\varphi(\mathbf{x}, \mathbf{y}) + \varphi(\mathbf{y}, \mathbf{x})] = \tfrac{1}{2}[\varphi(\mathbf{y}, \mathbf{x}) + \varphi(\mathbf{x}, \mathbf{y})] = \varphi_1(\mathbf{y}, \mathbf{x}),$$

while

$$\varphi_2(\mathbf{x}, \mathbf{y}) = \tfrac{1}{2}[\varphi(\mathbf{x}, \mathbf{y}) - \varphi(\mathbf{y}, \mathbf{x})] = -\tfrac{1}{2}[\varphi(\mathbf{y}, \mathbf{x}) - \varphi(\mathbf{x}, \mathbf{y})] = -\varphi_2(\mathbf{y}, \mathbf{x}).$$

The operation leading from the bilinear form φ to the bilinear form φ_1 is called *symmetrization* of φ, while the operation leading from φ to φ_2 is called *antisymmetrization* of φ. Obviously, φ can be represented as the sum

$$\varphi(\mathbf{x}, \mathbf{y}) = \varphi_1(\mathbf{x}, \mathbf{y}) + \varphi_2(\mathbf{x}, \mathbf{y}), \tag{3}$$

called the *decomposition of φ into its symmetric and antisymmetric parts.*

Next we express the tensors determined by the forms φ_1 and φ_2 in terms of the tensor determined by the form φ. Writing φ in component form, we have

$$\varphi(\mathbf{x}, \mathbf{y}) = a_{ij}x_iy_j,$$

where x_i and y_j are the components of the vectors $\mathbf{x}$ and $\mathbf{y}$, respectively. The forms φ_1 and φ_2 then become

$$\varphi_1(\mathbf{x}, \mathbf{y}) = \tfrac{1}{2}(a_{ij}x_iy_j + a_{ij}y_ix_j),$$
$$\varphi_2(\mathbf{x}, \mathbf{y}) = \tfrac{1}{2}(a_{ij}x_iy_j - a_{ij}y_ix_j).$$

But clearly

$$a_{ij}y_ix_j = a_{ji}y_jx_i = a_{ji}x_iy_j,$$

and hence

$$\varphi_1(\mathbf{x}, \mathbf{y}) = \tfrac{1}{2}(a_{ij} + a_{ji})x_iy_j,$$
$$\varphi_2(\mathbf{x}, \mathbf{y}) = \tfrac{1}{2}(a_{ij} - a_{ji})x_iy_j.$$

Let $a_{(ij)}$ denote the (symmetric) tensor determined by φ_1 and $a_{[ij]}$ the (antisymmetric) tensor determined by φ_2. Then

$$a_{(ij)} = \tfrac{1}{2}(a_{ij} + a_{ji}),$$
$$a_{[ij]} = \tfrac{1}{2}(a_{ij} - a_{ji}).$$

The operation leading from the tensor a_{ij} to the tensor $a_{(ij)}$ is called *symmetrization* of a_{ij}, while the operation leading from a_{ij} to $a_{[ij]}$ is called *antisymmetrization* of a_{ij}. Obviously

$$a_{ij} = a_{(ij)} + a_{[ij]},$$

in keeping with (3).

More generally, let $\varphi = \varphi(\mathbf{x}, \mathbf{y}, \mathbf{z}, \ldots, \mathbf{w})$ be a multilinear form of degree p. Then, in just the same way (give the details), we can define the operation of symmetrization (or antisymmetrization) of φ in two (given) arguments and corresponding operations on the tensor determined by φ. A somewhat more complicated problem is that of complete symmetrization (or antisymmetrization) of a multilinear form of degree $p > 2$. For example, to construct a form which is symmetric in all its arguments from a given trilinear form $\varphi = \varphi(\mathbf{x}, \mathbf{y}, \mathbf{z})$, we must carry out all possible permutations of the arguments of φ. There are precisely $3! = 6$ such permutations, and hence the desired symmetric trilinear form is just

$$\varphi_1(\mathbf{x}, \mathbf{y}, \mathbf{z}) = \tfrac{1}{6}[\varphi(\mathbf{x}, \mathbf{y}, \mathbf{z}) + \varphi(\mathbf{y}, \mathbf{z}, \mathbf{x}) + \varphi(\mathbf{z}, \mathbf{x}, \mathbf{y}) + \varphi(\mathbf{y}, \mathbf{x}, \mathbf{z}) + \varphi(\mathbf{z}, \mathbf{y}, \mathbf{x}) + \varphi(\mathbf{x}, \mathbf{z}, \mathbf{y})].$$

By the same token, the corresponding antisymmetric trilinear form is given by

$$\varphi_2(\mathbf{x}, \mathbf{y}, \mathbf{z}) = \tfrac{1}{6}[\varphi(\mathbf{x}, \mathbf{y}, \mathbf{z}) + \varphi(\mathbf{y}, \mathbf{z}, \mathbf{x}) + \varphi(\mathbf{z}, \mathbf{x}, \mathbf{y}) - \varphi(\mathbf{y}, \mathbf{x}, \mathbf{z}) - \varphi(\mathbf{z}, \mathbf{y}, \mathbf{x}) - \varphi(\mathbf{x}, \mathbf{z}, \mathbf{y})]$$

(verify the antisymmetry). The operations leading from the trilinear form to the forms φ_1 and φ_2 are again called symmetrization and antisymmetrization (of φ). Let a_{ijk} be the tensor determined by φ, $a_{(ijk)}$ the (symmetric) tensor determined by φ_1, and $a_{[ijk]}$ the (antisymmetric) tensor determined by φ_2. Then clearly

$$a_{(ijk)} = \tfrac{1}{6}(a_{ijk} + a_{jki} + a_{kij} + a_{jik} + a_{kji} + a_{ikj}),$$
$$a_{[ijk]} = \tfrac{1}{6}(a_{ijk} + a_{jki} + a_{kij} - a_{jik} - a_{kji} - a_{ikj}),$$

where the operations leading from the tensor a_{ijk} to the tensors $a_{(ijk)}$ and $a_{[ijk]}$ are once again called symmetrization and antisymmetrization (of a_{ijk}).

12.4. Next suppose we set $\mathbf{y} = \mathbf{x}$ in a bilinear form $\varphi = \varphi(\mathbf{x}, \mathbf{y})$. This gives a scalar function

$$\varphi = \varphi(\mathbf{x}, \mathbf{x})$$

of one vector argument, called a *quadratic form.* Obviously, every bilinear form $\varphi(\mathbf{x}, \mathbf{y})$ leads in this way to a unique quadratic form $\varphi(\mathbf{x}, \mathbf{x})$, but the same quadratic form may be "generated" by different bilinear forms. In fact, let $\varphi = \varphi(\mathbf{x}, \mathbf{y})$ be an arbitrary bilinear form, and let

$$\varphi_1(\mathbf{x}, \mathbf{y}) = \tfrac{1}{2}[\varphi(\mathbf{x}, \mathbf{y}) + \varphi(\mathbf{y}, \mathbf{x})]$$

be the bilinear form obtained by symmetrizing φ. Then

$$\varphi_1(\mathbf{x}, \mathbf{x}) = \tfrac{1}{2}[\varphi(\mathbf{x}, \mathbf{x}) + \varphi(\mathbf{x}, \mathbf{x})] = \varphi(\mathbf{x}, \mathbf{x}),$$

so that the two bilinear forms $\varphi(\mathbf{x}, \mathbf{y})$ and $\varphi_1(\mathbf{x}, \mathbf{y})$, which are in general distinct, generate the same quadratic form $\varphi(\mathbf{x}, \mathbf{x})$. Thus it can always be assumed that a given quadratic form $\varphi(\mathbf{x}, \mathbf{x})$ is obtained by setting $\mathbf{y} = \mathbf{x}$ in a *symmetric* bilinear form. This symmetric bilinear form is called the *polar* (*bilinear*) *form* of the given quadratic form $\varphi(\mathbf{x}, \mathbf{x})$. The polar form $\varphi(\mathbf{x}, \mathbf{y})$ is uniquely determined by its quadratic form. In fact,

$$\varphi(\mathbf{x} + \mathbf{y}, \mathbf{x} + \mathbf{y}) = \varphi(\mathbf{x}, \mathbf{x}) + \varphi(\mathbf{x}, \mathbf{y}) + \varphi(\mathbf{y}, \mathbf{x}) + \varphi(\mathbf{y}, \mathbf{y}).$$

But $\varphi(\mathbf{x}, \mathbf{y}) = \varphi(\mathbf{y}, \mathbf{x})$, and hence

$$\varphi(\mathbf{x}, \mathbf{y}) = \tfrac{1}{2}[\varphi(\mathbf{x} + \mathbf{y}, \mathbf{x} + \mathbf{y}) - \varphi(\mathbf{x}, \mathbf{x}) - \varphi(\mathbf{y}, \mathbf{y})].$$

Now let $\mathbf{e}_1, \mathbf{e}_2, \mathbf{e}_3$ be an orthonormal basis, and let $\varphi(\mathbf{x}, \mathbf{y})$ be the bilinear form polar to a given quadratic form $\varphi(\mathbf{x}, \mathbf{x})$. Then

$$\varphi(\mathbf{x}, \mathbf{y}) = a_{ij}x_i x_j,$$

where $a_{ij} = a_{ji}$ since $\varphi(\mathbf{x}, \mathbf{y})$ is symmetric, and hence

$$\varphi(\mathbf{x}, \mathbf{x}) = a_{ij}x_i x_j. \tag{4}$$

The expression (4) is a homogeneous polynomial of degree two in the components of the vector $\mathbf{x}$, with coefficients a_{ij} making up a symmetric tensor. In more detail, the quadratic form $\varphi(\mathbf{x}, \mathbf{x})$ is given by

$$\varphi(\mathbf{x}, \mathbf{x}) = a_{11}x_1^2 + a_{22}x_2^2 + a_{33}x_3^2 + 2a_{12}x_1x_2 + 2a_{13}x_1x_3 + 2a_{23}x_2x_3.$$

Conversely, any symmetric second-order tensor a_{ij} determines a unique quadratic form $\varphi(\mathbf{x}, \mathbf{x}) = a_{ij}x_i x_j$. Hence *there is a one-to-one correspondence between symmetric second-order tensors and quadratic forms.*

Example. The square of the length of the vector $\mathbf{x}$ is the quadratic form

$$|\mathbf{x}|^2 = x_i x_i = \delta_{ij} x_i x_j,$$

and the corresponding polar bilinear form is the scalar product

$$\mathbf{x} \cdot \mathbf{y} = x_i y_i = \delta_{ij} x_i x_j$$

of the vectors $\mathbf{x}$ and $\mathbf{y}$ (the symmetry of $\mathbf{x} \cdot \mathbf{y}$ has already been noted in Sec. 4).

Next suppose we set $\mathbf{y} = \mathbf{x}$, $\mathbf{z} = \mathbf{x}$ in a trilinear form $\varphi(\mathbf{x}, \mathbf{y}, \mathbf{z})$. This gives a scalar function $\varphi(\mathbf{x}, \mathbf{x}, \mathbf{x})$ of one vector argument, called a *cubic form*. Just as in the case of quadratic forms, it is easily proved that there is a one-to-one correspondence between cubic forms, symmetric trilinear forms and symmetric tensors of order three. Every cubic form φ can be written as

$$\varphi = a_{ijk} x_i x_j x_k$$

in terms of the components of the vector $\mathbf{x}$, where a_{ijk} is a symmetric tensor, or, in more detail, as

$$\begin{aligned}\varphi = a_{111}x_1^3 + a_{222}x_2^3 + a_{333}x_3^3 \\ + 3a_{112}x_1^2x_2 + 3a_{122}x_1x_2^2 + 3a_{113}x_1^2x_3 \\ + 3a_{133}x_1x_3^2 + 3a_{223}x_2^2x_3 + 3a_{233}x_2x_3^2 + 6a_{123}x_1x_2x_3,\end{aligned}$$

where the nine coefficients of the form coincide (apart from numerical factors) with the nine "essentially distinct" components of the symmetric tensor a_{ijk}.

Remark. More generally, by setting $\mathbf{y} = \mathbf{x}, \mathbf{z} = \mathbf{x}, \ldots, \mathbf{w} = \mathbf{x}$ in a multilinear form $\varphi(\mathbf{x}, \mathbf{y}, \mathbf{z}, \ldots, \mathbf{w})$ of degree p, we can construct a corresponding scalar function $\varphi(\mathbf{x}, \mathbf{x}, \mathbf{x}, \ldots, \mathbf{x})$ of one vector argument. Again it is easily proved that there is a one-to-one correspondence between the forms $\varphi(\mathbf{x}, \mathbf{x}, \mathbf{x}, \ldots, \mathbf{x})$, symmetric multilinear forms and symmetric tensors of order p.

12.5. Symmetric forms can be interpreted geometrically by introducing the concept of a "characteristic surface." Fixing an origin O in L_3, let $\mathbf{x} = \overrightarrow{OP}$ be the radius vector of a variable point $P \in L_3$. Then, given any symmetric second-order tensor a_{ij} with corresponding quadratic form $\varphi(\mathbf{x}, \mathbf{x}) = a_{ij}x_i x_j$, let S be the locus of all points P whose radius vectors $\mathbf{x}$ satisfy the condition

$$\varphi(\mathbf{x}, \mathbf{x}) = 1. \tag{5}$$

This locus is a surface S, called the *characteristic surface* of the tensor a_{ij}. In terms of the components of the vector $\mathbf{x}$ relative to some orthonormal basis $\mathbf{e}_1, \mathbf{e}_2, \mathbf{e}_3$, the equation of S takes the form

$$a_{ij} x_i x_j = 1. \tag{6}$$

It follows from the considerations of Sec. 7.7 that the characteristic surface of a symmetric second-order tensor is a central quadric surface whose center of symmetry coincides with the origin O.

Example 1. The characteristic surface of the unit tensor δ_{ij} has equation

$$\delta_{ij}x_ix_j = 1,$$

or equivalently,

$$x_1^2 + x_2^2 + x_3^2 = 1.$$

Thus the characteristic surface of the unit tensor is simply a sphere of unit radius.

Example 2. If $a_{ij} = a_ia_j$, then the characteristic surface of a_{ij} is just

$$a_ia_jx_ix_j = 1,$$

which can be written in the form

$$(a_ix_i)^2 = 1. \tag{7}$$

But (7) separates at once into two equations

$$a_ix_i = \pm 1,$$

and hence the characteristic surface of the tensor a_ia_j is a pair of parallel planes, symmetric with respect to the origin.

Returning to the characteristic surface (5) of an arbitrary symmetric tensor a_{ij}, let $\mathbf{x} = \overrightarrow{OP}$ be the radius vector of a variable point of the surface, and let $\mathbf{p}$ be a unit vector with the same direction as $\mathbf{x}$, so that

$$\mathbf{x} = x\mathbf{p}, \tag{8}$$

where $x = |\mathbf{x}|$ is the length of $\mathbf{x}$. Substituting (8) into (5) and using the linearity of the form φ in both its arguments, we get

$$x^2\varphi(\mathbf{p}, \mathbf{p}) = 1.$$

It follows that

$$\varphi(\mathbf{p}, \mathbf{p}) = \frac{1}{x^2},$$

i.e., *the value of a quadratic form* $\varphi(\mathbf{x}, \mathbf{x})$ *for* $\mathbf{x}$ *equal to a unit vector* $\mathbf{p}$ *is just the reciprocal of the square of the distance from the origin O to the point of the characteristic surface S in which the ray emanating from O with the direction of* $\mathbf{p}$ *intersects S.* In particular, if $\mathbf{p} = \mathbf{e}_i$ and if P_i is the point in which the ray emanating from O with the direction of $\mathbf{e}_i$ intersects S, then

$$\varphi(\mathbf{e}_i, \mathbf{e}_i) = \frac{1}{\alpha_i^2},$$

where $\alpha_i = |\overrightarrow{OP_i}|$. But $\varphi(\mathbf{e}_i, \mathbf{e}_i) = a_{ii}$ (no summation over i implied), and hence

$$a_{ii} = \frac{1}{\alpha_i^2}.$$

Remark 1. We can define the characteristic surface of a symmetric tensor of order greater than two in just the same way. For example, the characteristic surface S of a symmetric third-order tensor a_{ijk} has the equation

$$a_{ijk}x_ix_jx_k = 1. \tag{9}$$

Starting from (9), we can find the value of the cubic form

$$\varphi(\mathbf{x}, \mathbf{x}, \mathbf{x}) = a_{ijk}x_ix_jx_k$$

for $\mathbf{x}$ equal to a unit vector $\mathbf{p}$, namely

$$\varphi(\mathbf{p}, \mathbf{p}, \mathbf{p}) = \frac{1}{x^3},$$

where x is the distance from the origin O to the point of S in which the ray emanating from O with the direction of $\mathbf{p}$ intersects S. In particular,

$$a_{iii} = \varphi(\mathbf{e}_i, \mathbf{e}_i, \mathbf{e}_i) = \frac{1}{\alpha_i^3},$$

where α_i is the distance from the point O to the point in which the ray emanating from O with the direction of $\mathbf{e}_i$ intersects S.

Remark 2. Note that an equation of the form (6) or (9) can be used to define the characteristic surface of an arbitrary (not necessarily symmetric) tensor. But then the surface describes only the "symmetric part" of the tensor. For example, if a_{ij} is an arbitrary second-order tensor, then

$$a_{ij} = a_{(ij)} + a_{[ij]},$$

and equation (6) reduces to

$$a_{(ij)}x_ix_j = 1.$$

PROBLEMS

1. Prove that in the space L_3 every antisymmetric trilinear form $\varphi(\mathbf{x}, \mathbf{y}, \mathbf{z})$ differs from the scalar triple product $(\mathbf{x}, \mathbf{y}, \mathbf{z})$ by only a constant factor.

2. Prove that in the space L_3 every antisymmetric multilinear form of degree $p > 3$ is identically equal to zero.

3. State and prove theorems analogous to the assertions in Probs. 1 and 2 for the space L_n.

4. Prove that if the tensor a_{ijk} is symmetric in the indices i and j and antisymmetric in the indices j and k, then a_{ijk} vanishes.

5. Prove that if a_{ij} is a symmetric tensor and b_{ij} an antisymmetric tensor, then $a_{ij}b_{ij} = 0$.

6. Prove that if a tensor a_{ijk} is symmetric in its first two indices ($a_{ijk} = a_{jik}$) and if the relation

$$a_{ijk}x_ix_jx_k = 0$$

holds for every vector $\mathbf{x} = x_i\mathbf{e}_i$, then

$$a_{ijk} + a_{jki} + a_{kij} = 0.$$

7. Given a tensor a_{ij}, suppose

$$a_{ij}x_j = \alpha x_i$$

for every vector $\mathbf{x} = x_i\mathbf{e}_i$, where α is independent of $\mathbf{x}$. Prove that

$$a_{ij} = \alpha\delta_{ij}.$$

8. Given a tensor a_{ijkl}, suppose

$$a_{ijkl}x_iy_jx_ky_l = 0$$

for arbitrary vectors $\mathbf{x} = x_i\mathbf{e}_i$, $\mathbf{y} = y_j\mathbf{e}_j$. Prove that

$$a_{ijkl} + a_{jkli} + a_{klij} + a_{lijk} = 0.$$

Prove that

$$a_{ijkl} = 0$$

if, in addition,

$$a_{ijkl} + a_{jikl} = 0, \qquad a_{ijkl} + a_{ijlk} = 0, \qquad a_{(ijk)l} = 0.$$

9. Prove that every third-order tensor a_{ijk} can be written in the form

$$a_{ijk} = a_{(ijk)} + a_{[ijk]} + \tfrac{2}{3}(a_{[ij]k} + a_{[kj]i}) + \tfrac{2}{3}(a_{(ij)k} - a_{k(ij)}).$$

10. Prove that if a tensor a_{ijk} is symmetric in the indices i and j, then

$$a_{(ijk)} = \tfrac{1}{3}(a_{ijk} + a_{jki} + a_{kij}).$$

11. Prove that if a tensor a_{ijk} is antisymmetric in the indices i and j, then

$$a_{[ijk]} = \tfrac{1}{3}(a_{ijk} + a_{jki} + a_{kij}).$$

12. Decompose the tensor a_{ij} with matrix

$$(a_{ij}) = \begin{pmatrix} 2 & 3 & 2 \\ 5 & 7 & -2 \\ 4 & -4 & 0 \end{pmatrix}$$

into its symmetric part $b_{ij} = a_{(ij)}$ and antisymmetric part $c_{ij} = a_{[ij]}$. Then find

a) $c_{ij}a_{ij}$; b) $b_{ij}c_{ij}$; c) $c_{ij}\delta_{ij}$; d) $c_{ij}x_i$, where $\mathbf{x} = (2, 3, -4)$;
e) $c_{ij}x_ix_j$ (same $\mathbf{x}$); f) $b_{ij}\delta_{ij}$; g) $b_{ij}x_i$; h) $b_{ij}x_ix_j$.

13. Find the characteristic surface of the symmetric second-order tensor $a_{ij} = \lambda\delta_{ij}$. Do the same for the tensor $a_{ij} = \frac{1}{2}(a_ib_j + a_jb_i)$, where the vectors $\mathbf{a} = (a_1, a_2, a_3)$ and $\mathbf{b} = (b_1, b_2, b_3)$ are orthogonal.

14. If $n = 2$ the notion of a characteristic surface reduces to that of a "characteristic curve." Find the characteristic curves of the symmetric third-order tensors with the following components:

a) $a_{111} = a_{222} = 1$, $a_{112} = a_{122} = 0$;
b) $a_{111} = a_{222} = 0$, $a_{112} = a_{122} = \frac{1}{3}$;
c) $a_{111} = 1$, $a_{122} = -1$, $a_{112} = a_{222} = 0$.

(Sketch each curve after finding its equation.)

3

LINEAR TRANSFORMATIONS

13. Basic Concepts

13.1. So far we have considered *scalar* functions of one or several vector arguments in a linear space L. We now turn to the study of *vector* functions of a single vector argument, a topic of great importance in many branches of geometry, mechanics and physics. As we will see in Sec. 16, the most important of such functions, i.e., *linear* functions, are intimately related to second-order tensors.

Given a linear space L, by a *vector function* $\mathbf{A}$ defined on L we mean a rule associating a vector $\mathbf{u} = \mathbf{A}(\mathbf{x})$ with each vector $\mathbf{x} \in L$. A vector function $\mathbf{A}$ is said to be *linear* if

1) $\mathbf{A}(\mathbf{x} + \mathbf{y}) = \mathbf{A}(\mathbf{x}) + \mathbf{A}(\mathbf{y})$ for arbitrary vectors $\mathbf{x}$ and $\mathbf{y}$;
2) $\mathbf{A}(\alpha\mathbf{x}) = \alpha\mathbf{A}(\mathbf{x})$ for an arbitrary vector $\mathbf{x}$ and real number α,

A linear vector function is also called a *linear transformation* of the space L, or a *linear operator* (acting) in L. In writing vector functions, we will henceforth drop parentheses whenever this leads to no confusion, writing simply

$$\mathbf{u} = \mathbf{A}\mathbf{x}.$$

Geometrically, the first of the properties defining a linear vector function $\mathbf{A}$ means that $\mathbf{A}$ carries the diagonal of the parallelogram constructed on the vectors $\mathbf{x}$ and $\mathbf{y}$ into the diagonal of the parallelogram constructed on the vectors $\mathbf{u} = \mathbf{A}\mathbf{x}$ and $\mathbf{v} = \mathbf{A}\mathbf{y}$ (see Figure 5a). The second property means that if the length of the vector $\mathbf{x}$ is multiplied by a factor α, then so is the length of the vector $\mathbf{u} = \mathbf{A}\mathbf{x}$ (see Figure 5b). It follows that a linear transformation

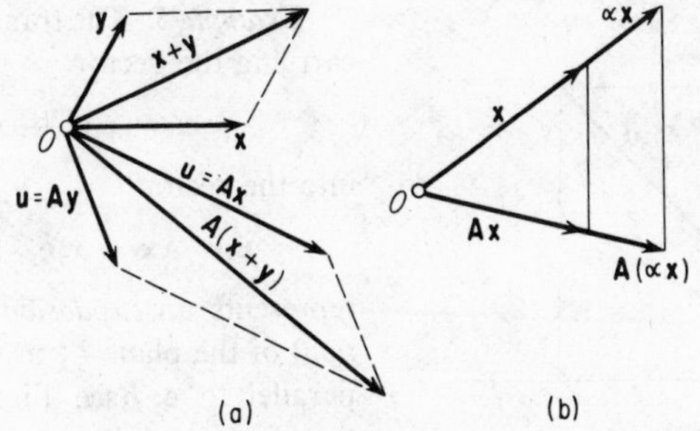

FIGURE 5

carries collinear vectors into collinear vectors and coplanar vectors into coplanar vectors (why?).

13.2. Next we give some examples of linear transformations.

Example 1. The transformation associating the vector $\mathbf{x}$ itself with every given vector $\mathbf{x}$ is obviously linear. This transformation, denoted by $\mathbf{E}$, is called the *identity* (or *unit*) *transformation.* Thus $\mathbf{Ex} = \mathbf{x}$ for all $\mathbf{x}$.

Example 2. The transformation associating the vector $\lambda\mathbf{x}$ (λ real) with every given vector $\mathbf{x}$ is also linear, since if $\mathbf{Ax} = \lambda\mathbf{x}$, then

$$\mathbf{A}(\mathbf{x} + \mathbf{y}) = \lambda(\mathbf{x} + \mathbf{y}) = \lambda\mathbf{x} + \lambda\mathbf{y} = \mathbf{Ax} + \mathbf{Ay},$$
$$\mathbf{A}(\alpha\mathbf{x}) = \lambda(\alpha\mathbf{x}) = \alpha(\lambda\mathbf{x}) = \alpha\mathbf{Ax}.$$

Geometrically, the transformation $\mathbf{Ax} = \lambda\mathbf{x}$ represents a *homogeneous expansion* (or *contraction*) of all vectors with the same expansion coefficient λ. Such a transformation is said to be *homothetic.* (If $\lambda < 0$, the vectors are reflected in the origin as well as expanded.)

Example 3. If $\lambda = 0$, the linear transformation considered in the preceding example associates the zero vector $\mathbf{0}$ with every vector $\mathbf{x}$. This transformation, denoted by $\mathbf{N}$, is called the *null* (or *zero*) *transformation.* Thus $\mathbf{Nx} = \mathbf{0}$ for all $\mathbf{x}$.

Example 4. The transformation

$$\mathbf{Ax} = \mathbf{x} + \mathbf{a} \qquad (\mathbf{a} \neq \mathbf{0})$$

is nonlinear, since

$$\mathbf{Ay} = \mathbf{y} + \mathbf{a},$$

and hence

$$\mathbf{A}(\mathbf{x} + \mathbf{y}) = \mathbf{x} + \mathbf{y} + \mathbf{a} \neq \mathbf{Ax} + \mathbf{Ay}.$$

We now consider some examples of linear transformations in the two-dimensional space L_2, equipped with an orthonormal basis $\mathbf{e}_1, \mathbf{e}_2$.

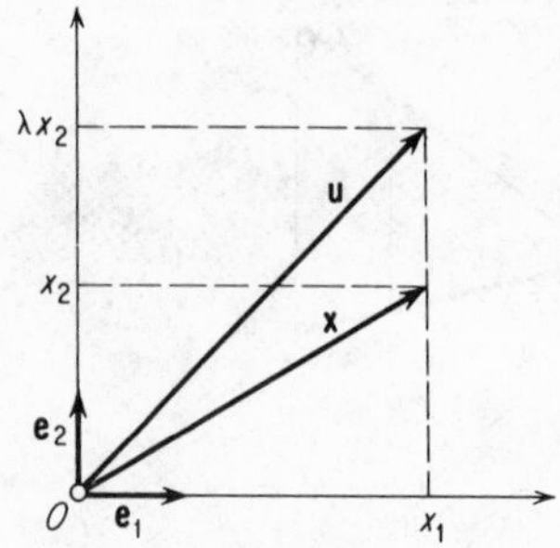

FIGURE 6

Example 5. The transformation **A** carrying the vector

$$\mathbf{x} = x_1\mathbf{e}_1 + x_2\mathbf{e}_2$$

into the vector

$$\mathbf{u} = \mathbf{A}\mathbf{x} = x_1\mathbf{e}_1 + \lambda x_2\mathbf{e}_2$$

represents an *expansion* (or *contraction*) of the plane L_2 in the direction parallel to $\mathbf{e}_2$ (see Figure 6). This transformation is linear, since

$$\begin{aligned}\mathbf{A}(\mathbf{x}+\mathbf{y}) &= (x_1+y_1)\mathbf{e}_1 + \lambda(x_2+y_2)\mathbf{e}_2\\ &= (x_1\mathbf{e}_1 + \lambda x_2\mathbf{e}_2) + (y_1\mathbf{e}_1 + \lambda y_2\mathbf{e}_2) = \mathbf{A}\mathbf{x} + \mathbf{A}\mathbf{y},\\ \mathbf{A}(\alpha x) &= (\alpha x_1)\mathbf{e}_1 + \lambda(\alpha x_2)\mathbf{e}_2 = \alpha(x_1\mathbf{e}_1 + \lambda x_2\mathbf{e}_2) = \alpha\mathbf{A}\mathbf{x}.\end{aligned}$$

Example 6. If $\lambda = 0$, the transformation just given reduces to the transformation

$$\mathbf{A}(x_1\mathbf{e}_1 + x_2\mathbf{e}_2) = x_1\mathbf{e}_1,$$

representing *projection* of the vector **x** onto the axis parallel to $\mathbf{e}_1$. Hence projection is a linear transformation.

Example 7. The transformation carrying every vector $\mathbf{x} \in L_2$ into the vector **u** obtained by rotating **x** through the angle θ (in the counterclockwise direction, say) is linear, as shown by the constructions in Figures 7a and 7b. Naturally, such a transformation is called a *rotation.*

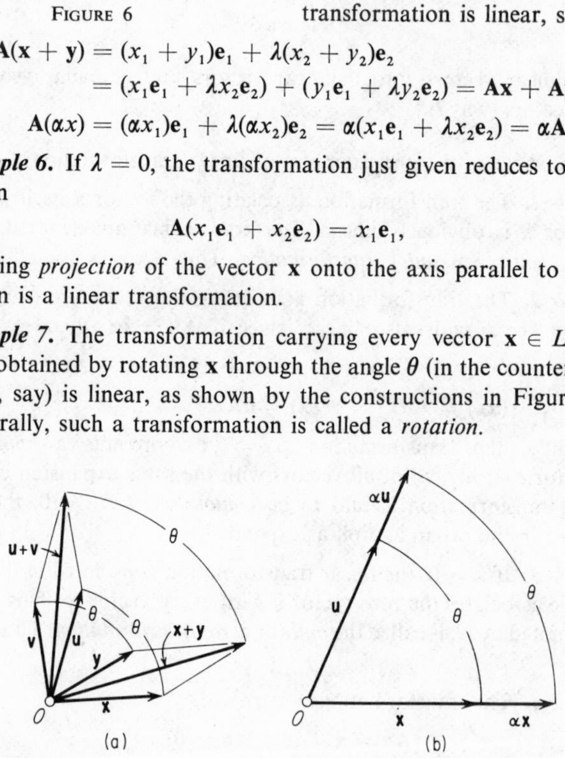

FIGURE 7

Example 8. The linearity of the transformation **A** carrying the vector

$$\mathbf{x} = x_1\mathbf{e}_1 + x_2\mathbf{e}_2$$

into the vector

$$\mathbf{u} = \mathbf{A}\mathbf{x} = (x_1 + kx_2)\mathbf{e}_1 + x_2\mathbf{e}_2$$

is proved in the same way as in Example 5. Note that **A** *shifts* the end of

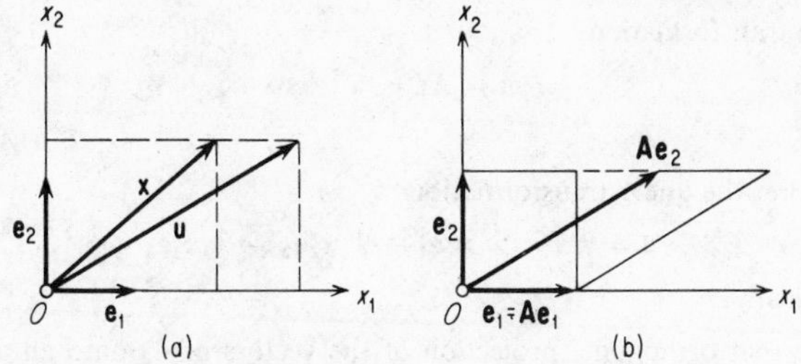

FIGURE 8

the vector **x** by an amount kx_2 along the line parallel to the x_1-axis (see Figure 8a), so that the square constructed on the vectors $\mathbf{e}_1$ and $\mathbf{e}_2$ goes into the parallelogram constructed on the vectors $\mathbf{e}_1$ and $\mathbf{e}_2 + k\mathbf{e}_1$ (see Figure 8b).

PROBLEMS

1. Prove that every linear transformation of a one-dimensional space is equivalent to multiplication of all vectors by the same number.

2. Let x_1 and x_2 be the components of an arbitrary vector **x** relative to a given basis $\mathbf{e}_1, \mathbf{e}_2$ in the plane L_2. Which of the following transformations are linear:

a) $\mathbf{u} = \mathbf{Ax} = -\mathbf{x}$;
b) $\mathbf{u} = \mathbf{Ax} = x_1\mathbf{e}_1 + x_1\mathbf{e}_2$;
c) $\mathbf{u} = \mathbf{Ax} = x_1\mathbf{e}_1 - 2x_2\mathbf{e}_2$;
d) $\mathbf{u} = \mathbf{Ax} = \lambda_1 x_1\mathbf{e}_1 + \lambda_2 x_2\mathbf{e}_2$;
e) $\mathbf{u} = \mathbf{Ax} = x_1^2\mathbf{e}_1$?

Interpret the linear transformations geometrically.

3. Write the transformation corresponding to expansion (or contraction) of the plane L_2 in the direction perpendicular to $\mathbf{e}_2$.

4. Suppose the basis vectors $\mathbf{e}_1$ and $\mathbf{e}_2$ in Example 5 are nonorthogonal. Prove that the corresponding transformation is linear, and interpret it geometrically.

5. Which of the following transformations of the space L_3 are linear:†

a) $\mathbf{u} = \mathbf{Ax} = (\mathbf{a}\cdot\mathbf{x})\mathbf{a}$;
b) $\mathbf{u} = \mathbf{Ax} = (\mathbf{a}\cdot\mathbf{x})\mathbf{x}$;
c) $\mathbf{u} = \mathbf{Ax} = \mathbf{a}$;
d) $\mathbf{u} = \mathbf{Ax} = x_1\mathbf{e}_1 + x_2\mathbf{e}_2$;
e) $\mathbf{u} = \mathbf{Ax} = x_1\mathbf{e}_1 - x_2\mathbf{e}_2 - 2x_3\mathbf{e}_3$;
f) $\mathbf{u} = \mathbf{Ax} = x_1\mathbf{e}_1 + x_2\mathbf{e}_2 + \lambda x_3\mathbf{e}_3$;
g) $\mathbf{u} = \mathbf{Ax} = x_2^3\mathbf{e}_2 + x_3\mathbf{e}_3$?

† In a)–c), **a** is a fixed nonzero vector, while in d)–g), x_1, x_2, x_3 are the components of an arbitrary vector **x** with respect to some orthonormal basis $\mathbf{e}_1, \mathbf{e}_2, \mathbf{e}_3$ (similarly in Probs. 6 and 7).

6. Is the transformation

$$\mathbf{u} = \mathbf{A}\mathbf{x} = \mathbf{a} \times \mathbf{x}$$

linear?

7. Interpret the linear transformation

$$\mathbf{u} = \mathbf{A}\mathbf{x} = \lambda_1 x_1 \mathbf{e}_1 + \lambda_2 x_2 \mathbf{e}_2 + \lambda_3 x_3 \mathbf{e}_3$$

geometrically.

8. Prove that orthogonal projection of the vectors of L_3 onto an axis making equal angles with the axes of a rectangular coordinate system is a linear transformation.

9. Prove that rotation of L_3 through the angle $2\pi/3$ about the line with equation $x_1 = x_2 = x_3$ relative to an orthonormal basis $\mathbf{e}_1, \mathbf{e}_2, \mathbf{e}_3$ is a linear transformation.

10. Prove that the operation of differentiation is linear in the space of all polynomials of degree not exceeding n.

11. Prove the linearity of the following transformations, defined on the space $C[a, b]$ of all functions continuous in the interval $[a, b]$:

a) $g(t) = \mathbf{A}f(t) = tf(t)$;

b) $g(t) = \mathbf{A}f(t) = f(t)\varphi(t)$, where $\varphi(t)$ is a fixed function continuous in $[a, b]$;

c) $g(t) = \mathbf{A}f(t) = \int_a^b H(t, s)f(s)\,ds$, where $H(t, s)$ is a fixed function continuous in both arguments.

12. Which of the transformations in Prob. 11 are linear in the space of all polynomials of degree not exceeding n?

14. The Matrix of a Linear Transformation and Its Determinant

14.1. Let $\mathbf{x}$ be an arbitrary vector in L_3, with expansion

$$\mathbf{x} = x_i\mathbf{e}_i = x_1\mathbf{e}_1 + x_2\mathbf{e}_2 + x_3\mathbf{e}_3$$

relative to a given orthonormal basis $\mathbf{e}_1, \mathbf{e}_2, \mathbf{e}_3$, and let

$$\mathbf{u} = \mathbf{A}\mathbf{x}$$

be a linear transformation of L_3, where $\mathbf{u}$ has the expansion

$$\mathbf{u} = u_i\mathbf{e}_i = u_1\mathbf{e}_1 + u_2\mathbf{e}_2 + u_3\mathbf{e}_3$$

relative to $\mathbf{e}_1, \mathbf{e}_2, \mathbf{e}_3$. We now find the relation between the components of the vector $\mathbf{u}$ and those of the original vector $\mathbf{x}$. Since the transformation $\mathbf{A}$ is linear, we have

$$\mathbf{A}\mathbf{x} = \mathbf{A}(x_1\mathbf{e}_1 + x_2\mathbf{e}_2 + x_3\mathbf{e}_3) = x_1\mathbf{A}\mathbf{e}_1 + x_2\mathbf{A}\mathbf{e}_2 + x_3\mathbf{A}\mathbf{e}_3. \tag{1}$$

Suppose that relative to the basis $\mathbf{e}_1, \mathbf{e}_2, \mathbf{e}_3$ the vectors $\mathbf{A}\mathbf{e}_1, \mathbf{A}\mathbf{e}_2, \mathbf{A}\mathbf{e}_3$ have the expansions

$$\begin{aligned}\mathbf{Ae}_1 &= a_{11}\mathbf{e}_1 + a_{21}\mathbf{e}_2 + a_{31}\mathbf{e}_3,\\ \mathbf{Ae}_2 &= a_{12}\mathbf{e}_1 + a_{22}\mathbf{e}_2 + a_{32}\mathbf{e}_3,\\ \mathbf{Ae}_3 &= a_{13}\mathbf{e}_1 + a_{23}\mathbf{e}_2 + a_{33}\mathbf{e}_3,\end{aligned} \tag{2}$$

or more concisely

$$\mathbf{Ae}_i = a_{ji}\mathbf{e}_j.$$

Then, substituting (2) into (1), we get

$$\begin{aligned}\mathbf{Ax} = (a_{11}x_1 + a_{12}x_2 + a_{13}x_3)\mathbf{e}_1 &+ (a_{21}x_1 + a_{22}x_2 + a_{23}x_3)\mathbf{e}_2\\ &+ (a_{31}x_1 + a_{32}x_2 + a_{33}x_3)\mathbf{e}_3,\end{aligned}$$

or more concisely

$$\mathbf{Ax} = a_{ij}x_j\mathbf{e}_i.$$

But $\mathbf{u} = \mathbf{Ax}$, and hence the components of $\mathbf{u}$ are just

$$\begin{aligned}u_1 &= a_{11}x_1 + a_{12}x_2 + a_{13}x_3,\\ u_2 &= a_{21}x_1 + a_{22}x_2 + a_{23}x_3,\\ u_3 &= a_{31}x_1 + a_{32}x_2 + a_{33}x_3,\end{aligned} \tag{3}$$

or briefly

$$u_i = a_{ij}x_j.$$

These formulas allow us to determine the components of the vector $\mathbf{u}$ obtained by subjecting the original vector $\mathbf{x}$ to the linear transformation $\mathbf{A}$. Note that the components of $\mathbf{u}$ are *homogeneous linear* expressions in the components of $\mathbf{x}$.

The coefficients of the formulas (3) relating the components of $\mathbf{u}$ and $\mathbf{x}$ can be written in the form of a matrix†

$$A = \begin{pmatrix} a_{11} & a_{12} & a_{13}\\ a_{21} & a_{22} & a_{23}\\ a_{31} & a_{32} & a_{33}\end{pmatrix},$$

called the *matrix of the linear transformation* $\mathbf{A}$. Note that A is a square matrix, with three rows and three columns. Thus we have proved that *to every linear transformation* $\mathbf{A}$ *of the space* L_3 *there corresponds a unique square matrix of order three* (relative to a given orthonormal basis in L_3). Conversely *to every square matrix A of order three there corresponds a unique linear transformation* (relative to the given basis). In fact, we need only use the matrix A to construct the vector function $\mathbf{u} = \mathbf{Ax}$ defined by the formulas (3), noting that the linearity of the vector function follows from the linearity and homogeneity of (3). Thus finally, *there is a one-to-one correspondence*

† Note that if an operator is denoted by a boldface Roman letter (like $\mathbf{A}$), then the matrix of the operator is denoted by the corresponding lightface Italic letter (like A).

between linear transformations of the space L_3 and square matrices of order three (relative to a given basis).

Remark 1. Consider a linear transformation $\mathbf{u} = \mathbf{Ax}$ of the plane L_2. Choosing a basis $\mathbf{e}_1$, $\mathbf{e}_2$ in L_2, we have

$$u_1 = a_{11}x_1 + a_{12}x_2,$$
$$u_2 = a_{21}x_1 + a_{22}x_2,$$

where

$$\mathbf{Ae}_1 = a_{11}\mathbf{e}_1 + a_{21}\mathbf{e}_2,$$
$$\mathbf{Ae}_2 = a_{12}\mathbf{e}_1 + a_{22}\mathbf{e}_2.$$

Hence any linear transformation $\mathbf{A}$ of the plane L_2 is described by a square matrix

$$A = \begin{pmatrix} a_{11} & a_{12} \\ a_{21} & a_{22} \end{pmatrix}$$

of order two.

Remark 2. More generally, consider a linear transformation $\mathbf{u} = \mathbf{Ax}$ of the n-dimensional space L_n. Choosing a basis $\mathbf{e}_1, \mathbf{e}_2, \ldots, \mathbf{e}_n$ in L_n, we have

$$\begin{aligned} u_1 &= a_{11}x_1 + a_{12}x_2 + \cdots + a_{1n}x_n, \\ u_2 &= a_{21}x_1 + a_{22}x_2 + \cdots + a_{2n}x_n, \\ &\cdot \quad \cdot \quad \cdot \quad \cdot \quad \cdot \quad \cdot \quad \cdot \quad \cdot \quad \cdot \\ u_n &= a_{n1}x_1 + a_{n2}x_2 + \cdots + a_{nn}x_n, \end{aligned}$$

where

$$\begin{aligned} \mathbf{Ae}_1 &= a_{11}\mathbf{e}_1 + a_{21}\mathbf{e}_2 + \cdots + a_{n1}\mathbf{e}_n, \\ \mathbf{Ae}_2 &= a_{12}\mathbf{e}_1 + a_{22}\mathbf{e}_2 + \cdots + a_{n2}\mathbf{e}_n, \\ &\cdot \quad \cdot \quad \cdot \quad \cdot \quad \cdot \quad \cdot \quad \cdot \quad \cdot \quad \cdot \\ \mathbf{Ae}_n &= a_{1n}\mathbf{e}_1 + a_{2n}\mathbf{e}_2 + \cdots + a_{nn}\mathbf{e}_n. \end{aligned}$$

Hence any linear transformation $\mathbf{A}$ of L_n is described by a square matrix

$$A = (a_{ij}) = \begin{pmatrix} a_{11} & a_{12} & \ldots & a_{1n} \\ a_{21} & a_{22} & \ldots & a_{2n} \\ \cdot & \cdot & \ldots & \cdot \\ a_{n1} & a_{n2} & \ldots & a_{nn} \end{pmatrix}$$

of order n.

14.2. We now give a number of examples illustrating the above considerations.

Example 1. If $\mathbf{E}$ is the identity transformation, then

$$\mathbf{u} = \mathbf{Ex} = \mathbf{x},$$

and hence $u_i = x_i$, so that the matrix of $\mathbf{E}$ has the form

$$E = \begin{pmatrix} 1 & 0 & 0 \\ 0 & 1 & 0 \\ 0 & 0 & 1 \end{pmatrix}$$

in every basis. More concisely

$$E = (\delta_{ij}),$$

in terms of the Kronecker delta

$$\delta_{ij} = \begin{cases} 1 \text{ if } i = j, \\ 0 \text{ if } i \neq j. \end{cases}$$

The matrix E is called the *unit matrix*.

Example 2. Under the homothetic transformation

$$\mathbf{u} = \mathbf{A}\mathbf{x} = \lambda\mathbf{x},$$

the components of the vectors $\mathbf{u}$ and $\mathbf{x}$ are related by the formula $u_i = \lambda x_i$, so that the matrix of $\mathbf{A}$ has the form

$$A = \begin{pmatrix} \lambda & 0 & 0 \\ 0 & \lambda & 0 \\ 0 & 0 & \lambda \end{pmatrix}$$

in every basis, or more concisely

$$A = (\lambda\delta_{ij}).$$

Example 3. Under the null transformation

$$\mathbf{u} = \mathbf{N}\mathbf{x} \equiv \mathbf{0},$$

we have $u_i = 0$, and hence the matrix N of the null transformation consists entirely of zeros:

$$N = \begin{pmatrix} 0 & 0 & 0 \\ 0 & 0 & 0 \\ 0 & 0 & 0 \end{pmatrix}.$$

The matrix N is called the *null* (or *zero*) *matrix*.

Remark. More generally, each of the matrices E, A and N considered in Examples 1–3 has the same form in n-dimensional space as in three-dimensional space. For example, in n-dimensional space E is the square matrix

$$E = \begin{pmatrix} 1 & 0 & \dots & 0 \\ 0 & 1 & \dots & 0 \\ \cdot & \cdot & \dots & \cdot \\ 0 & 0 & \dots & 1 \end{pmatrix}$$

of order n.

Example 4. The transformation $\mathbf{A}$ carrying the vector $\mathbf{x} = x_1\mathbf{e}_1 + x_2\mathbf{e}_2$ into the vector $\mathbf{u} = x_1\mathbf{e}_1 + \lambda x_2\mathbf{e}_2$ represents an expansion (or contraction) of the plane L_2 in the direction parallel to $\mathbf{e}_2$ (recall Example 5, p. 66). Here

$$u_1 = x_1, \qquad u_2 = \lambda x_2,$$

so that the matrix of the transformation is just

$$A = \begin{pmatrix} 1 & 0 \\ 0 & \lambda \end{pmatrix}.$$

Example 5. If $\lambda = 0$, the transformation considered in the preceding example reduces to projection onto the axis parallel to $\mathbf{e}_1$, with matrix

$$A = \begin{pmatrix} 1 & 0 \\ 0 & 0 \end{pmatrix}.$$

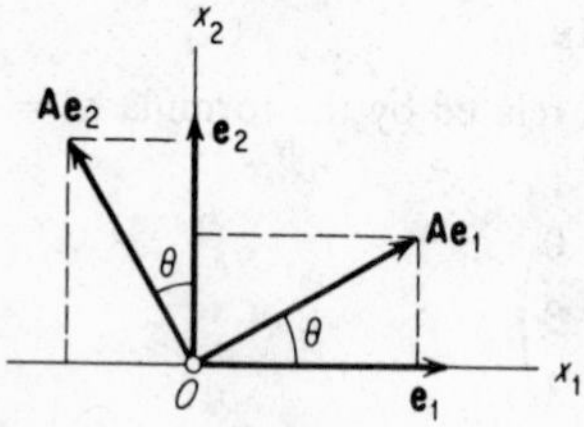

FIGURE 9

Example 6. Let $\mathbf{A}$ be the transformation which rotates the plane L_2 through the angle θ (recall Example 7, p. 66). Then

$$\mathbf{Ae}_1 = \mathbf{e}_1 \cos\theta + \mathbf{e}_2 \sin\theta,$$
$$\mathbf{Ae}_2 = -\mathbf{e}_1 \sin\theta + \mathbf{e}_2 \cos\theta$$

(see Figure 9). It follows that

$$\begin{aligned}\mathbf{u} = \mathbf{Ax} = \mathbf{A}(x_1\mathbf{e}_1 + x_2\mathbf{e}_2) &= x_1\mathbf{Ae}_1 + x_2\mathbf{Ae}_2 \\ &= (x_1 \cos\theta - x_2 \sin\theta)\mathbf{e}_1 + (x_1 \sin\theta + x_2 \cos\theta)\mathbf{e}_2,\end{aligned}$$

and hence

$$u_1 = x_1 \cos\theta - x_2 \sin\theta,$$
$$u_2 = x_1 \sin\theta + x_2 \cos\theta.$$

Therefore the matrix of the transformation $\mathbf{A}$ is just

$$A = \begin{pmatrix} \cos\theta & -\sin\theta \\ \sin\theta & \cos\theta \end{pmatrix}.$$

Example 7. The transformation $\mathbf{A}$ carrying the vector $\mathbf{x} = x_1\mathbf{e}_1 + x_2\mathbf{e}_2$ into the vector $\mathbf{u} = (x_1 + kx_2)\mathbf{e}_1 + x_2\mathbf{e}_2$ represents a *shift* of the plane L_2 in the direction parallel to $\mathbf{e}_1$ (recall Example 8, p. 66). Here

$$u_1 = x_1 + kx_2, \qquad u_2 = x_2,$$

so that the matrix of the transformation is simply

$$A = \begin{pmatrix} 1 & k \\ 0 & 1 \end{pmatrix}.$$

Example 8. Consider the transformation $\mathbf{A}$ of the plane L_2 carrying the vector $\mathbf{x} = x_1\mathbf{e}_1 + x_2\mathbf{e}_2$ into the vector $\mathbf{u} = \lambda_1 x_1\mathbf{e}_1 + \lambda_2 x_2\mathbf{e}_2$. Then $\mathbf{A}$ is linear (why?), with matrix

$$A = \begin{pmatrix} \lambda_1 & 0 \\ 0 & \lambda_2 \end{pmatrix}.$$

Geometrically, this transformation represents a combination of two simultaneous expansions (or contractions) of the plane along two perpendicular axes $\mathbf{e}_1$ and $\mathbf{e}_2$, with expansion coefficients λ_1 and λ_2, respectively. If either of these expansion coefficients is negative, say λ_1, then the λ_1-fold expansion is accompanied by reflection in the line of $\mathbf{e}_2$.

Example 9. In just the same way, consider the transformation $\mathbf{A}$ of the space L_3 carrying the vector $\mathbf{x} = x_1\mathbf{e}_1 + x_2\mathbf{e}_2 + x_3\mathbf{e}_3$ into the vector $\mathbf{u} = \lambda_1 x_1 \mathbf{e}_1 + \lambda_2 x_2 \mathbf{e}_2 + \lambda_3 x_3 \mathbf{e}_3$. Then $\mathbf{A}$ is linear, with matrix

$$A = \begin{pmatrix} \lambda_1 & 0 & 0 \\ 0 & \lambda_2 & 0 \\ 0 & 0 & \lambda_3 \end{pmatrix}.$$

Geometrically, this transformation consists of three simultaneous expansions (or contractions) of space along three perpendicular axes $\mathbf{e}_1$, $\mathbf{e}_2$ and $\mathbf{e}_3$, with expansion coefficients λ_1, λ_2 and λ_3, respectively. A matrix like A, with all of its elements equal to zero except those on the main diagonal,† is called a *diagonal matrix*. In particular, if $\lambda_1 = \lambda_2 = \lambda_3$, then $\mathbf{A}$ reduces to a homothetic transformation, while if $\lambda_1 = \lambda_2 \neq \lambda_3$, then $\mathbf{A}$ is a homothetic transformation only in the plane of the vectors $\mathbf{e}_1$ and $\mathbf{e}_2$.

14.3. Let $\mathbf{u} = \mathbf{A}\mathbf{x}$ be a linear transformation in the space L_3 equipped with an orthonormal basis $\mathbf{e}_1, \mathbf{e}_2, \mathbf{e}_3$. Then $\mathbf{A}$ carries the basis vectors into the vectors

$$\mathbf{a}_i = \mathbf{A}\mathbf{e}_i = a_{1i}\mathbf{e}_1 + a_{2i}\mathbf{e}_2 + a_{3i}\mathbf{e}_3,$$

where, as we have seen, the components of the vectors $\mathbf{a}_i$ make up the columns of the matrix of the transformation $\mathbf{A}$. Under the transformation $\mathbf{A}$ the vector $\mathbf{x} = x_i\mathbf{e}_i$ goes into the vector

$$\mathbf{u} = \mathbf{A}\mathbf{x} = x_i\mathbf{A}\mathbf{e}_i = x_i\mathbf{a}_i.$$

Thus the expansion of $\mathbf{u}$ with respect to the vectors $\mathbf{a}_i$ has the same coefficients as the expansion of the original vector $\mathbf{x}$ with respect to the basis vectors $\mathbf{e}_i$.

Now consider the unit cube constructed on the basis vectors $\mathbf{e}_1, \mathbf{e}_2, \mathbf{e}_3$. Then the "oriented volume" V_e of this cube equals ± 1, depending on whether the triple of vectors $\mathbf{e}_1, \mathbf{e}_2, \mathbf{e}_3$ is right-handed or left-handed. In terms of the quantity ϵ introduced in Sec. 5.1, we have

$$V_e = \epsilon.$$

† Naturally, some (or all) of the diagonal elements may also equal zero.

Under the transformation **A** the cube constructed on the vectors $\mathbf{e}_1, \mathbf{e}_2, \mathbf{e}_3$ goes into a parallelepiped (in general, nonrectangular) constructed on the vectors $\mathbf{a}_1, \mathbf{a}_2, \mathbf{a}_3$. The oriented volume V_a of this parallelepiped equals the scalar triple product of the vectors $\mathbf{a}_1, \mathbf{a}_2, \mathbf{a}_3$, i.e.,

$$V_a = (\mathbf{a}_1, \mathbf{a}_2, \mathbf{a}_3).$$

Using the representation of $(\mathbf{a}_1, \mathbf{a}_2, \mathbf{a}_3)$ as a determinant (see Sec. 5.2), we have

$$V_a = \epsilon \begin{vmatrix} a_{11} & a_{21} & a_{31} \\ a_{12} & a_{22} & a_{32} \\ a_{13} & a_{23} & a_{33} \end{vmatrix}. \tag{4}$$

The determinant in (4) differs from the determinant of the matrix of the transformation **A** (see p. 69) in that rows and columns have been interchanged. But this has no effect on the value of a determinant, and hence

$$V_a = \epsilon\,|A|,$$

where $|A|$ denotes the determinant of the matrix A.

Next consider an arbitrary parallelepiped constructed on given vectors $\mathbf{x}_1, \mathbf{x}_2, \mathbf{x}_3$. Under the transformation **A** this parallelepiped goes into the parallelepiped constructed on the vectors

$$\mathbf{u}_1 = \mathbf{A}\mathbf{x}_1, \qquad \mathbf{u}_2 = \mathbf{A}\mathbf{x}_2, \qquad \mathbf{u}_3 = \mathbf{A}\mathbf{x}_3,$$

where, as just noted, the expansions of the vectors $\mathbf{u}_i$ with respect to the vectors $\mathbf{a}_i$ have the same coefficients as the expansions of the original vectors $\mathbf{x}_i$ with respect to the basis vectors $\mathbf{e}_i$. Hence, if V_x denotes the (oriented) volume of the parallelepiped constructed on the vectors $\mathbf{x}_1, \mathbf{x}_2, \mathbf{x}_3$, while V_u denotes the volume of the parallelepiped constructed on the vectors $\mathbf{u}_1, \mathbf{u}_2, \mathbf{u}_3$, we have

$$\frac{V_u}{V_a} = \frac{V_x}{V_e},$$

and therefore

$$\frac{V_u}{V_x} = |A|.$$

Thus *the determinant of the matrix of a linear transformation measures the "magnification"*† *of volumes as a result of the transformation.* If $|A| > 0$, the oriented volumes V_u and V_x have the same sign, and hence the transformation **A** preserves the orientation of vectors. On the other hand, if $|A| < 0$, the transformation **A** changes the orientation of vectors into the opposite orientation.

Suppose now that $|A| = 0$. Then

$$(\mathbf{a}_1, \mathbf{a}_2, \mathbf{a}_3) = 0,$$

† Here the word "magnification" is used in a general sense, comprising both "stretching" and "shrinking."

and the vectors $\mathbf{a}_1, \mathbf{a}_2, \mathbf{a}_3$ are linearly dependent. Suppose $\mathbf{a}_1, \mathbf{a}_2, \mathbf{a}_3$ are noncollinear, and let Π denote the plane determined by these vectors. Then every vector $\mathbf{x} = x_i\mathbf{e}_i$ goes into a vector $\mathbf{u} = x_i\mathbf{a}_i$ lying in the plane Π, i.e., the linear transformation $\mathbf{A}$ carries every vector of space into a vector lying in Π. If, however, the vectors $\mathbf{a}_1, \mathbf{a}_2, \mathbf{a}_3$ are collinear, all lying on a line l, then $\mathbf{A}$ carries every vector of space into a vector lying on l. Finally, if $\mathbf{a}_1 = \mathbf{a}_2 = \mathbf{a}_3 = \mathbf{0}$, then $\mathbf{A}$ carries every vector $\mathbf{x} \in L_3$ into the zero vector.

A linear transformation $\mathbf{A}$ or the corresponding matrix A is said to be *singular* if the determinant $|A|$ vanishes. The "degree of singularity" of $\mathbf{A}$ differs from case to case (as we have just seen) and can be made precise by introducing a new concept, namely "rank." By the *rank* of the matrix

$$A = \begin{pmatrix} a_{11} & a_{12} & a_{13} \\ a_{21} & a_{22} & a_{23} \\ a_{31} & a_{32} & a_{33} \end{pmatrix}$$

we mean the largest of the orders of the nonzero determinants contained in A.† If $|A| \neq 0$, the rank of the matrix A equals three. If $|A| = 0$ and the vectors $\mathbf{a}_1, \mathbf{a}_2, \mathbf{a}_3$ are noncollinear, then A must contain a nonzero determinant of order two (since at least two of its columns are nonproportional), i.e., the rank of A equals two. If $|A| = 0$ and the vectors $\mathbf{a}_1, \mathbf{a}_2, \mathbf{a}_3$ are collinear, then all the second-order determinants contained in A vanish and the rank of A equals one (here, of course, we assume that at least one of the vectors $\mathbf{a}_1, \mathbf{a}_2, \mathbf{a}_3$ is nonzero!). Finally, the only matrix of rank zero is the null matrix N.

Conversely, suppose $|A| = 0$ and let r be the rank of A. Then the matrix A contains two linearly independent columns if $r = 2$ and one linearly independent column (i.e., one nonzero column) if $r = 1$, while every column consists entirely of zeros if $r = 0$. Correspondingly, two of the vectors $\mathbf{a}_1, \mathbf{a}_2, \mathbf{a}_3$ are linearly independent if $r = 2$ and one of the vectors $\mathbf{a}_1, \mathbf{a}_2, \mathbf{a}_3$ is linearly independent (nonzero) if $r = 1$, while all three vectors $\mathbf{a}_1, \mathbf{a}_2, \mathbf{a}_3$ vanish if $r = 0$.

The preceding considerations are summarized in the following

THEOREM. *Let* $\mathbf{A}$ *be a linear transformation of the space* L_3, *and let* r $(0 \leq r \leq 3)$ *be the rank of the matrix of* $\mathbf{A}$. *Then* $\mathbf{A}$ *maps the whole space* L_3 *into the r-dimensional linear space* L_r.

Example 1. Consider the projection of the space L_3 onto the plane perpendicular to the vector $\mathbf{e}_3$, i.e., the linear transformation $\mathbf{A}$ carrying the vector $\mathbf{x} = x_1\mathbf{e}_1 + x_2\mathbf{e}_2 + x_3\mathbf{e}_3$ into the vector $\mathbf{u} = \mathbf{A}\mathbf{x} = x_1\mathbf{e}_1 + x_2\mathbf{e}_2$. Then

$$u_1 = x_1, \qquad u_2 = x_2, \qquad u_3 = 0,$$

† More exactly, made up of the elements at the intersections of k rows and k columns of A $(1 \leq k \leq 3)$.

and the transformation **A** has the matrix

$$\begin{pmatrix} 1 & 0 & 0 \\ 0 & 1 & 0 \\ 0 & 0 & 0 \end{pmatrix}$$

of rank two. A more general transformation with matrix of rank two is given by

$$\mathbf{u} = \mathbf{A}\mathbf{x} = \mathbf{a}_1(\mathbf{b}_1 \cdot \mathbf{x}) + \mathbf{a}_2(\mathbf{b}_2 \cdot \mathbf{x}), \tag{5}$$

where both pairs of vectors $\mathbf{a}_1$, $\mathbf{a}_2$ and $\mathbf{b}_1$, $\mathbf{b}_2$ are noncollinear. This transformation projects the whole space L_3 onto the plane determined by the vectors $\mathbf{a}_1$ and $\mathbf{a}_2$.

Example 2. Given a unit vector $\mathbf{e}_0$, the transformation

$$\mathbf{u} = \mathbf{A}\mathbf{x} = \mathbf{e}_0(\mathbf{e}_0 \cdot \mathbf{x}),$$

projecting every vector $\mathbf{x} \in L_3$ onto the axis with direction specified by $\mathbf{e}_0$, is a transformation whose matrix is of rank one. A more general transformation with matrix of rank one is given by

$$\mathbf{u} = \mathbf{A}\mathbf{x} = \mathbf{a}(\mathbf{b} \cdot \mathbf{x}). \tag{6}$$

PROBLEMS

1. Find the matrix (relative to the basis $\mathbf{e}_1, \mathbf{e}_2$) of the linear transformations of the plane L_2 considered in Probs. 2 and 3, p. 67.

2. Prove that under expansion (or contraction) of the plane L_2 (cf. Example 5, p. 66), a circle with center at the origin goes into an ellipse, while an equilateral hyperbola with the coordinate axes as its axes goes into a general hyperbola.

3. Find the matrices (relative to the basis $\mathbf{e}_1, \mathbf{e}_2, \mathbf{e}_3$) of the linear transformations of the space L_3 considered in Probs. 5–9, pp. 67–68.

4. Prove that expansion (or contraction) of the space L_3 along the x_3-axis (cf. Prob. 5f, p. 67) carries a sphere with center at the origin into an ellipsoid of revolution and the ellipsoid of revolution

$$\frac{x_1^2}{a_1^2} + \frac{x_2^2}{a_2^2} + \frac{x_3^2}{a_2^2} = 1,$$

into a general ellipsoid. Prove that the same transformation carries the hyperboloid of revolution

$$-\frac{x_1^2}{a_1^2} + \frac{x_2^2}{a_2^2} + \frac{x_3^2}{a_2^2} = \pm 1$$

of one or two sheets into a general hyperboloid of one or two sheets.

5. Let **A** be the "differentiation operator" in the space of all polynomials $P(t)$ of degree not exceeding n, i.e., the operator such that $\mathbf{A}P(t) = P'(t)$. Find the matrix of **A** relative to the following bases:

a) $1, t, t^2, \ldots, t^n$;

b) $1, t - a, \dfrac{(t-a)^2}{2!}, \ldots, \dfrac{(t-a)^n}{n!}$.

6. Prove that there exists a unique linear transformation **C** of the space L_3 carrying three linearly independent vectors $\mathbf{a}_1, \mathbf{a}_2, \mathbf{a}_3$ into three (not necessarily linearly independent) vectors $\mathbf{b}_1, \mathbf{b}_2, \mathbf{b}_3$. Find the matrix C of this transformation relative to a given orthonormal basis $\mathbf{e}_1, \mathbf{e}_2, \mathbf{e}_3$.

7. Write the matrix of the linear transformation **C** of the space L_3 carrying the vectors

$$\mathbf{a}_1 = (2, 3, 5), \qquad \mathbf{a}_2 = (0, 1, 2), \qquad \mathbf{a}_3 = (1, 0, 0)$$

into the vectors

$$\mathbf{b}_1 = (1, 1, 1), \qquad \mathbf{b}_2 = (1, 1, -1), \qquad \mathbf{b}_3 = (2, 1, 2),$$

respectively.

8. Describe geometrically the linear transformations of the space L_3 with the following matrices relative to an orthonormal basis $\mathbf{e}_1, \mathbf{e}_2, \mathbf{e}_3$:

$$\text{a)} \begin{pmatrix} -1 & 0 & 0 \\ 0 & 1 & 0 \\ 0 & 0 & 1 \end{pmatrix}; \quad \text{b)} \begin{pmatrix} 1 & 0 & 0 \\ 0 & \lambda & 0 \\ 0 & 0 & \lambda \end{pmatrix}; \quad \text{c)} \begin{pmatrix} 1 & 0 & 0 \\ 0 & 0 & 0 \\ 0 & 0 & 1 \end{pmatrix}; \quad \text{d)} \begin{pmatrix} 0 & 0 & 0 \\ 0 & 1 & 0 \\ 0 & 0 & 0 \end{pmatrix}.$$

9. Prove that a rotation of the space L_3 through an angle α about the axis defined by the unit vector $\boldsymbol{\omega}$ is the linear transformation given by the formula

$$\mathbf{u} = \mathbf{A}\mathbf{x} = (\mathbf{x}\cdot\boldsymbol{\omega})\boldsymbol{\omega} + [\mathbf{x} - (\mathbf{x}\cdot\boldsymbol{\omega})\boldsymbol{\omega}] \cos\alpha + \boldsymbol{\omega} \times \mathbf{x} \sin\alpha.$$

Find the matrix of this transformation in the basis $\mathbf{e}_1, \mathbf{e}_2, \mathbf{e}_3$ if $\boldsymbol{\omega} = \omega_i \mathbf{e}_i$.

10. State and prove the analogue for the plane L_2 of the theorem on p. 75.

11. Which of the linear transformations considered in Sec. 13 and in Probs. 2–10, pp. 67–68 are nonsingular and which are singular? Find the rank of the matrix of each singular transformation.

12. Verify the linearity of the transformations (5) and (6), write the corresponding matrices, and verify that the matrices have ranks 2 and 1, respectively.

13. Describe geometrically the linear transformations of the plane L_2 and of the space L_3 with the following matrices in some orthonormal basis:

$$\text{a)} \begin{pmatrix} 1 & 1 \\ 2 & 2 \end{pmatrix}; \quad \text{b)} \begin{pmatrix} 1 & 1 & 2 \\ 2 & 2 & 1 \\ 3 & 3 & 3 \end{pmatrix}; \quad \text{c)} \begin{pmatrix} 1 & 2 & 3 \\ 2 & 4 & 6 \\ 3 & 6 & 9 \end{pmatrix}.$$

Find the rank of each matrix.

14. Prove that a linear transformation **A** is nonsingular if and only if

a) $\mathbf{A}\mathbf{x} = \mathbf{0}$ implies $\mathbf{x} = \mathbf{0}$;

b) **A** carries three linearly independent vectors of the space L_3 into three linearly independent vectors;

c) **A** is a one-to-one mapping, i.e., $\mathbf{x} \neq \mathbf{y}$ implies $\mathbf{Ax} \neq \mathbf{Ay}$;

d) **A** maps the space L_3 into the whole space L_3, i.e., given any vector $\mathbf{y} \in L_3$, there is a vector $\mathbf{x} \in L_3$ such that $\mathbf{Ax} = \mathbf{y}$.

15. Prove that the image and inverse image under a linear transformation **A** of a linear subspace L of the space L_3 are both linear subspaces.†

16. By the *null space* of a linear transformation **A** defined on a linear space L we mean the set of vectors in L which **A** carries into the zero vector **0**. The dimension of the null space of **A** is called the *defect* of **A**. By the *range* of the transformation **A** we mean the image under **A** of the whole space L. The dimension of the range of **A** is called the *rank* of **A**. Prove that

a) The rank of the transformation **A** equals the rank of its matrix;

b) The sum of the rank and the defect of **A** equals the dimension of L;

c) The defect of the transformation **A** equals the defect of its matrix, the defect of a matrix of order n and rank r being defined as the number $n - r$.

17. Prove that the linear transformation **A** is nonsingular if and only if

a) The null space of **A** contains only the zero vector, i.e., the defect of **A** equals zero;

b) The range of **A** coincides with the whole space L, i.e., the rank of **A** equals the dimension of L.

18. Find the null space, range, defect and rank of each of the transformations of the spaces L_2 and L_3 with the following matrices (in some orthonormal basis):

a) $\begin{pmatrix} a & 0 \\ 1 & 0 \end{pmatrix}$; b) $\begin{pmatrix} 1 & 0 & 0 \\ 0 & 0 & 0 \\ 0 & 0 & 3 \end{pmatrix}$; c) $\begin{pmatrix} 0 & 0 & 0 \\ 0 & 0 & 0 \\ 0 & 0 & 1 \end{pmatrix}$; d) $\begin{pmatrix} 0 & 0 & 0 \\ 0 & 0 & 1 \\ 1 & 0 & 0 \end{pmatrix}$.

19. Find the null space, range, defect and rank of the differentiation operator in the space of all polynomials $P(t)$ of degree not exceeding n.

15. Linear Transformations and Bilinear Forms

15.1. Let **x** and **y** be arbitrary vectors of the linear space L_3, and let **A** be a linear transformation of L_3. Consider the scalar product of the vector **x** and **u**, where $\mathbf{u} = \mathbf{Ay}$ is the result of applying the transformation **A** to the vector **y**. Then the expression‡

$$\varphi(\mathbf{x}, \mathbf{y}) = \mathbf{x} \cdot \mathbf{u} = (\mathbf{x}, \mathbf{Ay}) \tag{1}$$

is a scalar function of the vector arguments **x** and **y**. Clearly φ is a bilinear

---

† By the *image* of L under **A** we mean the set of all **y** such that $\mathbf{y} = \mathbf{Ax}$ for some $\mathbf{x} \in L$, while by the *inverse image* (synonymously, *preimage*) of L under **A** we mean the set of all **x** such that $\mathbf{Ax} = \mathbf{y}$ for some $\mathbf{y} \in L$.

‡ Here we use the alternative notation (. , .) for the scalar product (see p. 12).

form, since

$$\varphi(\mathbf{x}_1 + \mathbf{x}_2, \mathbf{y}) = (\mathbf{x}_1 + \mathbf{x}_2, \mathbf{A}\mathbf{y}) = (\mathbf{x}_1, \mathbf{A}\mathbf{y}) + (\mathbf{x}_2, \mathbf{A}\mathbf{y}) = \varphi(\mathbf{x}_1, \mathbf{y}) + \varphi(\mathbf{x}_2, \mathbf{y}),$$
$$\varphi(\mathbf{x}, \mathbf{y}_1 + \mathbf{y}_2) = (\mathbf{x}, \mathbf{A}(\mathbf{y}_1 + \mathbf{y}_2)) = (\mathbf{x}, \mathbf{A}\mathbf{y}_1) + (\mathbf{x}, \mathbf{A}\mathbf{y}_2) = \varphi(\mathbf{x}, \mathbf{y}_1) + \varphi(\mathbf{x}, \mathbf{y}_2),$$
$$\varphi(\lambda\mathbf{x}, \mathbf{y}) = (\lambda\mathbf{x}, \mathbf{A}\mathbf{y}) = \lambda(\mathbf{x}, \mathbf{A}\mathbf{y}) = \lambda\varphi(\mathbf{x}, \mathbf{y}),$$
$$\varphi(\mathbf{x}, \lambda\mathbf{y}) = (\mathbf{x}, \mathbf{A}\lambda\mathbf{y}) = (\mathbf{x}, \lambda\mathbf{A}\mathbf{y}) = \lambda(\mathbf{x}, \mathbf{A}\mathbf{y}) = \lambda\varphi(\mathbf{x}, \mathbf{y}).$$

THEOREM. *The matrix of the linear transformation* **A** *coincides with the coefficient matrix of the bilinear form* (1).

Proof. Let

$$\mathbf{x} = x_i\mathbf{e}_i, \qquad \mathbf{y} = y_i\mathbf{e}_i, \qquad \mathbf{u} = u_i\mathbf{e}_i$$

relative to an orthonormal basis $\mathbf{e}_1, \mathbf{e}_2, \mathbf{e}_3$ in L_3. Since $\mathbf{u} = \mathbf{A}\mathbf{y}$, we have

$$u_i = a_{ij}y_j,$$

where $A = (a_{ij})$ is the matrix of the transformation **A**. But then

$$\varphi(\mathbf{x}, \mathbf{y}) = x_iu_i = a_{ij}x_iy_j,$$

i.e., the elements of the matrix A are just the elements of the coefficient matrix of φ. ∎

COROLLARY. *A matrix* $A = (a_{ij})$ *is the matrix of a linear transformation* **A** *if and only if* a_{ij} *is a second-order tensor.*

Proof. If **A** is a linear transformation with matrix $A = (a_{ij})$, then A is the coefficient matrix of the bilinear form (1). Hence a_{ij} is a second-order tensor, by the definition on p. 46.

Conversely, let a_{ij} be a second-order tensor, and let x_i be the components of an arbitrary vector $\mathbf{x} \in L_3$. Then it follows from Sec. 11.5 that the numbers

$$u_i = a_{ij}x_j \qquad (i = 1, 2, 3) \tag{2}$$

are the components of a new vector **u**. The vector function

$$\mathbf{u} = \mathbf{A}(\mathbf{x}) = \mathbf{A}\mathbf{x}, \tag{2'}$$

equivalent to (2), is obviously linear, i.e., **A** is a linear transformation, in fact the transformation with matrix $A = (a_{ij})$. ∎

15.2. As noted in the remark on p. 50, the set of all tensors of a given order p forms a linear space of dimension 3^p. In particular, this applies to the case of second-order tensors ($p = 2$). Given two second-order tensors a_{ij} and b_{ij}, let **A** and **B** be the corresponding linear transformations, i.e., the transformations with matrices $A = (a_{ij})$ and $B = (b_{ij})$. Forming the sum $c_{ij} = a_{ij} + b_{ij}$ (itself a tensor), let **C** be the linear transformation with matrix $C = (c_{ij})$. Then **C** is called the *sum* of the transformations **A** and **B**, denoted by

$$\mathbf{C} = \mathbf{A} + \mathbf{B}.$$

Similarly, given a second-order tensor a_{ij} and a real number λ, form the product $d_{ij} = \lambda a_{ij}$ (again a tensor), and let $\mathbf{D}$ be the linear transformation with matrix $\mathbf{D} = (d_{ij})$. Then $\mathbf{D}$ is called the *product* of the transformation $\mathbf{A}$ with the number λ, denoted by

$$\mathbf{D} = \lambda\mathbf{A}.$$

It is easy to interpret the transformation $\mathbf{C}$ geometrically. Given any vector $\mathbf{x} \in L_3$, let

$$\mathbf{y} = \mathbf{Ax}, \qquad \mathbf{z} = \mathbf{Bx}, \qquad \mathbf{u} = \mathbf{Cx}.$$

Then

$$\mathbf{u} = \mathbf{y} + \mathbf{z}$$

(see Figure 10a), since

$$u_i = c_{ij}x_j = (a_{ij} + b_{ij})x_j = a_{ij}x_j + b_{ij}x_j = y_j + z_i,$$

i. e.,

$$(\mathbf{A} + \mathbf{B})\mathbf{x} = \mathbf{Ax} + \mathbf{Bx}.$$

In just the same way,

$$(\lambda\mathbf{A})\mathbf{x} = \lambda(\mathbf{Ax})$$

(see Figure 10b). Since the set of all second-order tensors is a linear space of dimension 9, the same is true of the set of all linear transformations of L_3.

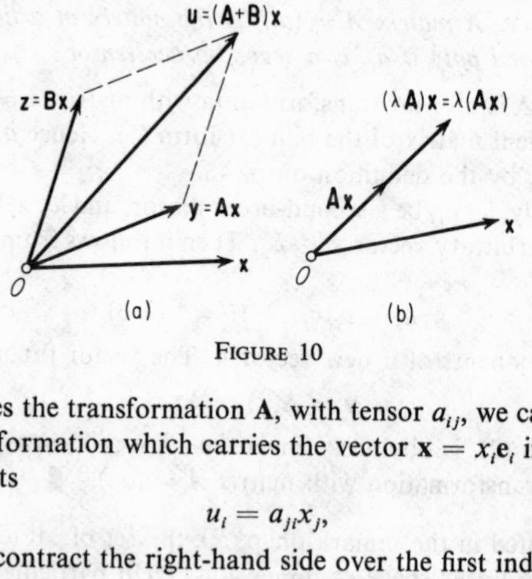

FIGURE 10

15.3. Besides the transformation $\mathbf{A}$, with tensor a_{ij}, we can also consider the linear transformation which carries the vector $\mathbf{x} = x_i\mathbf{e}_i$ into the vector $\mathbf{u}$ with components

$$u_i = a_{ji}x_j, \tag{3}$$

where we now contract the right-hand side over the first index of the tensor a_{ij} rather than over the second index as in (2). This transformation, denoted by the symbol $\mathbf{A}^*$, is called the *adjoint* of the transformation $\mathbf{A}$. Setting $a_{ij}^* = a_{ji}$, we can write (3) in the form

$$u_i = a_{ij}^* x_j. \tag{3'}$$

Thus the transformation $\mathbf{A}^*$ has the matrix $A^* = (a_{ij}^*)$ obtained by *transposing* the matrix $A = (a_{ij})$ of the original transformation $\mathbf{A}$, i.e., by interchanging rows and columns of A.

THEOREM. *If* **A** *is a linear transformation with adjoint* **A***, *then*

$$(\mathbf{x}, \mathbf{A}\mathbf{y}) = (\mathbf{y}, \mathbf{A}^*\mathbf{x}) \tag{4}$$

for arbitrary vectors **x** *and* **y**.

Proof. Consider the bilinear form

$$\varphi(\mathbf{x}, \mathbf{y}) = (x, \mathbf{A}\mathbf{y}) = a_{ij}x_i y_j, \tag{5}$$

where $A = (a_{ij})$ is the matrix of **A** and x_i, y_j are the components of the vectors **x**, **y** (relative to an underlying orthonormal basis $\mathbf{e}_1$, $\mathbf{e}_2$, $\mathbf{e}_3$). We can also write (5) as

$$\varphi(\mathbf{x}, \mathbf{y}) = y_j(a_{ij}x_i) = y_j u_j, \tag{6}$$

where

$$u_j = a_{ij}x_i. \tag{7}$$

But the vector **u** with components (7) is the result of applying the transformation **A*** to the vector $\mathbf{x} = x_i\mathbf{e}_i$, as we see at once by interchanging the indices i and j in (3). Therefore (6) takes the form

$$\varphi(\mathbf{x}, \mathbf{y}) = (\mathbf{y}, \mathbf{A}^*\mathbf{x}).$$

Comparing this with (5), we immediately get (4). ∎

15.4. A linear transformation **A** is called *symmetric* (synonymously, *self-adjoint*) if it coincides with its own adjoint **A***.

THEOREM. *A linear transformation* **A** *is symmetric if and only if the bilinear form*

$$\varphi(\mathbf{x}, \mathbf{y}) = (\mathbf{x}, \mathbf{A}\mathbf{y})$$

associated with **A** *is symmetric.*

Proof. Suppose **A** is symmetric, so that $\mathbf{A} = \mathbf{A}^*$. Then

$$(\mathbf{x}, \mathbf{A}\mathbf{y}) = (\mathbf{y}, \mathbf{A}^*\mathbf{x}) = (\mathbf{y}, \mathbf{A}\mathbf{x}),$$

and hence φ is symmetric, i.e.,

$$\varphi(\mathbf{x}, \mathbf{y}) = \varphi(\mathbf{y}, \mathbf{x}). \tag{8}$$

Conversely, suppose φ is symmetric, so that (8) holds. Then

$$(\mathbf{x}, \mathbf{A}\mathbf{y}) = (\mathbf{y}, \mathbf{A}\mathbf{x}), \tag{9}$$

and comparing (9) with (4), we get

$$(\mathbf{y}, \mathbf{A}\mathbf{x}) = (\mathbf{y}, \mathbf{A}^*\mathbf{x}). \tag{10}$$

Since (10) holds for arbitrary **y**, we must have

$$\mathbf{A}\mathbf{x} = \mathbf{A}^*\mathbf{x}. \tag{11}$$

But (11) holds in turn for arbitrary **x**, and hence $\mathbf{A} = \mathbf{A}^*$, i.e., **A** is symmetric. ∎

COROLLARY. *The matrix $A = (a_{ij})$ of a linear transformation* **A** *is symmetric (i.e., $a_{ij} = a_{ji}$) if and only if the transformation* **A** *is symmetric.*

Proof. The transformation **A** is symmetric if and only if the bilinear form $\varphi(\mathbf{x}, \mathbf{y}) = (\mathbf{x}, \mathbf{A}\mathbf{y})$ is symmetric. But, by Sec. 12.1, φ is symmetric if and only if $a_{ij} = a_{ji}$. ∎

Remark. Comparing this corollary with the corollary on p. 79, we see that a matrix $A = (a_{ij})$ is the matrix of a *symmetric* linear transformation **A** if and only if a_{ij} is a *symmetric* second-order tensor, i.e., there is a one-to-one correspondence between symmetric linear transformations and symmetric second-order tensors. It follows from the italicized assertion on p. 59 that *there is a one-to-one correspondence between symmetric linear transformations and quadratic forms.*†

Next consider the characteristic surface S of the tensor of a symmetric linear transformation **A** (briefly, the characteristic surface of the transformation **A**). According to Sec. 12.5, the equation of S is just

$$a_{ij}x_ix_j = 1,$$

or equivalently

$$(\mathbf{x}, \mathbf{A}\mathbf{x}) = 1.$$

Given any vector **x**, let P be the point such that the vector $\overrightarrow{OP}$, joining the origin to the point P, has the direction of **x** (see Figure 11). Then the vector $\mathbf{u} = \mathbf{A}\mathbf{x}$ has the direction of the normal to S at the point P. In fact, any normal to the surface with equation $\varphi(x_1, x_2, x_3) = c$ in a rectangular coordinate system is proportional to the vector with components‡

$$\frac{\partial \varphi}{\partial x_i}.$$

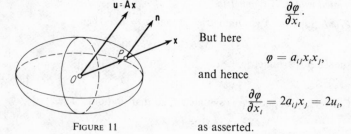

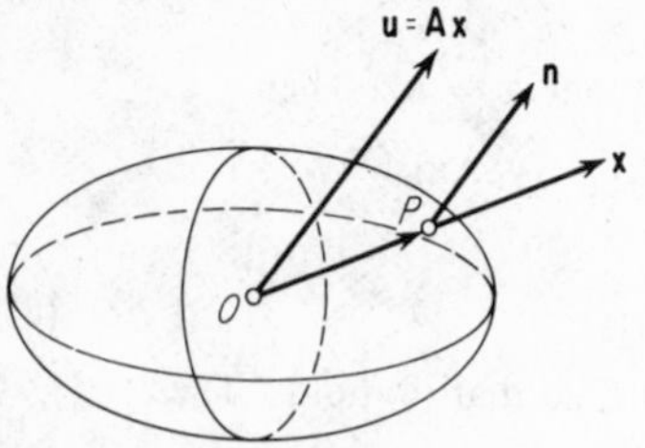

FIGURE 11

But here

$$\varphi = a_{ij}x_ix_j,$$

and hence

$$\frac{\partial \varphi}{\partial x_i} = 2a_{ij}x_j = 2u_i,$$

as asserted.

15.5. A linear transformation **A** is called *antisymmetric* if it is the negative of its own adjoint, i.e., if

$$\mathbf{A} = -\mathbf{A}^*.$$

---

† The quadratic form φ associated with the transformation **A** is, of course, just $\varphi(\mathbf{x}, \mathbf{x}) = (\mathbf{x}, \mathbf{A}\mathbf{x})$.

‡ See, e.g., R. A. Silverman, *Modern Calculus and Analytic Geometry*, The Macmillan Co., New York (1969), p. 732.

Just as in the case of a symmetric transformation, it can be shown that a linear transformation is antisymmetric if and only if the bilinear form $\varphi(\mathbf{x}, \mathbf{y}) = (\mathbf{x}, \mathbf{A}\mathbf{y}) = a_{ij}x_iy_j$ is antisymmetric, i.e., satisfies the condition

$$a_{ij} = -a_{ji}$$

(so that, in particular, $a_{ii} = 0$).

Now consider the vector $\mathbf{a} = a_i\mathbf{e}_i$, where

$$a_i = -\tfrac{1}{2}\epsilon_{ijk}a_{jk}. \tag{12}$$

Recalling the meaning of ϵ_{ijk} from p. 17, we have

$$a_1 = -\epsilon a_{23}, \qquad a_2 = -\epsilon a_{31}, \qquad a_3 = -\epsilon a_{12},$$

where the quantity ϵ equals $+1$ if the basis $\mathbf{e}_1, \mathbf{e}_2, \mathbf{e}_3$ is right-handed and -1 if the basis is left-handed. Therefore the matrix of an antisymmetric linear transformation can be written in the form

$$(a_{ij}) = \epsilon \begin{pmatrix} 0 & -a_3 & a_2 \\ a_3 & 0 & -a_1 \\ -a_2 & a_1 & 0 \end{pmatrix}.$$

Any antisymmetric linear transformation $\mathbf{A}$ can be written in the form

$$\mathbf{A}\mathbf{x} = \mathbf{a} \times \mathbf{x},$$

where $\mathbf{a}$ is the vector with components (12). In fact, if $\mathbf{u} = \mathbf{A}\mathbf{x}$, then

$$u_1 = a_{1j}x_j = \epsilon(-a_3x_2 + a_2x_3),$$
$$u_2 = a_{2j}x_j = \epsilon(a_3x_1 - a_1x_3),$$
$$u_3 = a_{3j}x_j = \epsilon(-a_2x_1 + a_1x_2).$$

But the expressions on the right are just the components of the vector product $\mathbf{a} \times \mathbf{x}$ (see p. 18).

15.6. Finally, we find the bilinear forms corresponding to some of the linear transformations considered in the preceding sections.

Example 1. The bilinear form corresponding to the identity transformation $\mathbf{E}\mathbf{x} = \mathbf{x}$ is just

$$\varphi(\mathbf{x}, \mathbf{y}) = (\mathbf{x}, \mathbf{E}\mathbf{y}) = (\mathbf{x}, \mathbf{y}), \tag{13}$$

i.e., the scalar product of the vectors $\mathbf{x}$ and $\mathbf{y}$. Since the form φ is symmetric, so is the transformation $\mathbf{E}$. The corresponding quadratic form is

$$\varphi(\mathbf{x}, \mathbf{x}) = (\mathbf{x}, \mathbf{E}\mathbf{x}) = (\mathbf{x}, \mathbf{x}) = |\mathbf{x}|^2,$$

and hence the characteristic surface of $\mathbf{E}$ is the unit sphere

$$|\mathbf{x}|^2 = 1.$$

Example 2. The bilinear form corresponding to the homothetic transformation

$$\mathbf{A}\mathbf{x} = \lambda\mathbf{x} \tag{14}$$

is

$$\varphi(\mathbf{x}, \mathbf{y}) = (\mathbf{x}, \lambda\mathbf{y}) = \lambda(\mathbf{x}, \mathbf{y}),$$

differing from (13) only by the factor λ. The form φ is symmetric, like the transformation (14) itself. The matrix of φ (and of **A**) is just $(\lambda\delta_{ij})$. The quadratic form corresponding to the transformation (14) is

$$\varphi(\mathbf{x}, \mathbf{x}) = (\mathbf{x}, \mathbf{A}\mathbf{x}) = \lambda|\mathbf{x}|^2,$$

and hence the characteristic surface of **A** is the sphere

$$\lambda|\mathbf{x}|^2 = 1$$

of radius

$$R = \frac{1}{\sqrt{\lambda}}.$$

(For this reason, the tensor $\lambda\delta_{ij}$ is often called *spherical.*) Note that the coefficient λ may be negative, in which case the characteristic surface is a sphere of "imaginary radius."

Example 3. Let **A** be the transformation carrying the vector $x = x_i\mathbf{e}_i$ into the vector

$$\mathbf{u} = \mathbf{A}\mathbf{x} = \lambda_1 x_1 \mathbf{e}_1 + \lambda_2 x_2 \mathbf{e}_2 + \lambda_3 x_3 \mathbf{e}_3.$$

Then the bilinear form corresponding to **A** is

$$\varphi(\mathbf{x}, \mathbf{y}) = (\mathbf{x}, \mathbf{A}\mathbf{y}) = \lambda_1 x_1 y_1 + \lambda_2 x_2 y_2 + \lambda_3 x_3 y_3.$$

The form φ is symmetric, and so is the transformation **A**. In fact, the matrix of **A** is diagonal (recall Example 9, p. 73), and hence obviously symmetric. The quadratic form corresponding to **A** is

$$\varphi(\mathbf{x}, \mathbf{x}) = \lambda_1 x_1^2 + \lambda_2 x_2^2 + \lambda_3 x_3^2,$$

while the characteristic surface of **A** is

$$\lambda_1 x_1^2 + \lambda_2 x_2^2 + \lambda_3 x_3^2 = 1.$$

This is the equation of a central quadric surface, with the coordinate axes as its axes of symmetry. If all the "expansion coefficients" λ_i are positive, the surface is an *ellipsoid.* If two of the numbers λ_i are positive and one is negative, the surface is a *hyperboloid of one sheet,* while if one of the numbers λ_i is positive and two are negative, the surface is a *hyperboloid of two sheets.* Finally, if all the λ_i are negative, the characteristic surface is an "imaginary ellipsoid." If any two of the numbers λ_i are equal, the characteristic surface is a *surface of revolution*, while if $\lambda_1 = \lambda_2 = \lambda_3$, the surface reduces to a sphere.

Example 4. The transformation **A** rotating the plane L_2 about the origin through the angle θ in the counterclockwise direction has the matrix

$$A = \begin{pmatrix} \cos\theta & -\sin\theta \\ \sin\theta & \cos\theta \end{pmatrix},$$

as shown in Example 6, p. 72. The bilinear form corresponding to this transformation is

$$\begin{aligned}(\mathbf{x}, \mathbf{y}) = (\mathbf{x}, \mathbf{A}\mathbf{y}) &= x_1 y_1 \cos\theta - x_1 y_2 \sin\theta + x_2 y_1 \sin\theta + x_2 y_2 \cos\theta \\ &= (x_1 y_1 + x_2 y_2)\cos\theta - (x_1 y_2 - x_2 y_1)\sin\theta.\end{aligned}$$

This bilinear form is no longer symmetric, and hence the transformation $\mathbf{A}^*$ has the matrix

$$A^* = \begin{pmatrix} \cos\theta & \sin\theta \\ -\sin\theta & \cos\theta \end{pmatrix},$$

and corresponds geometrically to a rotation about O through the angle $-\theta$.

Example 5. Let $\mathbf{A}$ be the transformation of the plane L_2 considered in Example 7, p. 72, with matrix

$$A = \begin{pmatrix} 1 & \lambda \\ 0 & 1 \end{pmatrix}.$$

This transformation is nonsymmetric, and the same is true of the associated bilinear form

$$\varphi(\mathbf{x}, \mathbf{y}) = (\mathbf{x}, \mathbf{A}\mathbf{y}) = x_1 y_1 + \lambda x_1 y_2 + x_2 y_2.$$

The transformation $\mathbf{A}^*$ adjoint to $\mathbf{A}$ has the matrix

$$A^* = \begin{pmatrix} 1 & 0 \\ \lambda & 1 \end{pmatrix},$$

and corresponds geometrically to a "shift" like $\mathbf{A}$, but in the direction of $\mathbf{e}_1$ rather than of $\mathbf{e}_2$.

PROBLEMS

1. Prove the symmetry of the following linear transformations of the plane L_2 (x_1 and x_2 are the components of an arbitrary vector $\mathbf{x} \in L_2$):

a) $\mathbf{u} = \mathbf{A}\mathbf{x} = x_1\mathbf{e}_1$;
b) $\mathbf{u} = \mathbf{A}\mathbf{x} = -\mathbf{x}$;
c) $\mathbf{u} = \mathbf{A}\mathbf{x} = x_1\mathbf{e}_1 - x_2\mathbf{e}_2$;
d) $\mathbf{u} = \mathbf{A}\mathbf{x} = x_1\mathbf{e}_1 + 3x_2\mathbf{e}_2$;
e) $\mathbf{u} = \mathbf{A}\mathbf{x} = x_1\mathbf{e}_1 + \lambda x_2\mathbf{e}_2$;
f) $\mathbf{u} = \mathbf{A}\mathbf{x} = \lambda_1 x_1\mathbf{e}_1 + \lambda_2 x_2\mathbf{e}_2$.

Find the corresponding quadratic $\varphi = \varphi(\mathbf{x}, \mathbf{x})$ and characteristic curves.

2. Do the same for the following linear transformations of the space L_3 (x_1, x_2 and x_3 are the components of an arbitrary vector $\mathbf{x} \in L_3$, while $\mathbf{a} = (a_1, a_2, a_3)$ and $\mathbf{b} = (b_1, b_2, b_3)$ are a pair of fixed orthogonal vectors):

a) $\mathbf{u} = \mathbf{A}\mathbf{x} = x_2\mathbf{e}_2$;
b) $\mathbf{u} = \mathbf{A}\mathbf{x} = x_1\mathbf{e}_1 + x_2\mathbf{e}_2$;
c) $\mathbf{u} = \mathbf{A}\mathbf{x} = x_1\mathbf{e}_1 + x_2\mathbf{e}_2 - x_3\mathbf{e}_3$;

d) $\mathbf{u} = \mathbf{A}\mathbf{x} = -x_1\mathbf{e}_1 + 2x_2\mathbf{e}_2 - x_3\mathbf{e}_3$;
e) $\mathbf{u} = \mathbf{A}\mathbf{x} = (\mathbf{a}\cdot\mathbf{x})\mathbf{a}$;
f) $\mathbf{u} = \mathbf{A}\mathbf{x} = (\mathbf{a}\cdot\mathbf{x})\mathbf{a} + (\mathbf{b}\cdot\mathbf{x})\mathbf{b}$.

3. Find the adjoint of each of the following linear transformations of the space L_3:
a) $\mathbf{u} = \mathbf{A}\mathbf{x} = (x_1 + 2x_2)\mathbf{e}_1 + x_2\mathbf{e}_2 + x_3\mathbf{e}_3$;
b) $\mathbf{u} = \mathbf{A}\mathbf{x} = -x_2\mathbf{e}_1 + x_1\mathbf{e}_2 + x_3\mathbf{e}_3$;
c) $\mathbf{u} = \mathbf{A}\mathbf{x} = (\mathbf{a}\cdot\mathbf{x})\mathbf{b}$;
d) $\mathbf{u} = \mathbf{A}\mathbf{x} = (\mathbf{a}_1\cdot\mathbf{x})\mathbf{b}_1 + (\mathbf{a}_2\cdot\mathbf{x})\mathbf{b}_2$;
e) $\mathbf{u} = \mathbf{A}\mathbf{x} = \mathbf{a} \times \mathbf{x}$.
Express each transformation as a sum of a symmetric part and an antisymmetric part.

4. Prove the following properties of the adjoint of a linear transformation (or the transpose of a matrix):
a) $(\mathbf{A}^*)^* = \mathbf{A}$;
b) $(\mathbf{A} + \mathbf{B})^* = \mathbf{A}^* + \mathbf{B}^*$;
c) $(\lambda\mathbf{A})^* = \lambda\mathbf{A}^*$;
d) $\mathbf{E}^* = \mathbf{E}$.

5. The matrix B of a linear transformation $\mathbf{B}$ coincides in some basis with the matrix A^* of the transformation $\mathbf{A}^*$ adjoint to the linear transformation $\mathbf{A}$. Is the same true in every basis?

6. Prove directly that addition of linear transformations (and matrices) and multiplication of transformations by real numbers have the following properties:
a) $\mathbf{A} + \mathbf{B} = \mathbf{B} + \mathbf{A}$;
b) $\mathbf{A} + (\mathbf{B} + \mathbf{C}) = (\mathbf{A} + \mathbf{B}) + \mathbf{C}$;
c) $\lambda(\mathbf{A} + \mathbf{B}) = \lambda\mathbf{A} + \lambda\mathbf{B}$;
d) $(\lambda + \mu)\mathbf{A} = \lambda\mathbf{A} + \mu\mathbf{A}$;
e) $(\lambda\mathbf{A} + \mu\mathbf{B})^* = \lambda\mathbf{A}^* + \mu\mathbf{B}^*$.

7. Prove that the operation of reflection in a plane Π in the direction of a line l is a symmetric linear transformation if and only if the line l is perpendicular to the plane Π.

8. Let the scalar product of the functions f and g in the space $C[a, b]$ be defined by the formula

$$(f, g) = \int_a^b f(t)g(t)\,dt,$$

as in Prob. 6, p. 15. Prove that
a) The linear transformation corresponding to multiplication by t (see Prob. 11a, p. 68) is symmetric;
b) The linear transformation

$$\mathbf{A}f(t) = \int_a^b H(t, s)f(s)\,ds$$

(see Prob. 11c, p. 68), where $H(t, s)$ is a fixed function continuous in both arguments such that $H(t, s) = H(s, t)$, is symmetric;

c) The linear transformation

$$\mathbf{A}f(t) = f'(t)$$

is antisymmetric if $f(a) = f(b) = 0$;

d) The linear transformation

$$\mathbf{A}f(t) = f''(t)$$

is symmetric if $f(a) = f(b), f'(a) = f'(b)$.

16. Multiplication of Linear Transformations and Matrices

16.1. Let $\mathbf{A}$ and $\mathbf{B}$ be two linear transformations of the space L_3. Suppose we subject an arbitrary vector $\mathbf{x}$ to the transformation $\mathbf{A}$, obtaining a vector $\mathbf{y} = \mathbf{Ax}$, and afterwards subject $\mathbf{y}$ to the transformation $\mathbf{B}$, obtaining a third vector $\mathbf{z} = \mathbf{By}$. Then $\mathbf{z}$ can be regarded as a vector function of the vector argument $\mathbf{x}$:

$$\mathbf{z} = \mathbf{Cx} = \mathbf{B}(\mathbf{Ax}).$$

Clearly, $\mathbf{C}$ is a linear transformation, since

$$\mathbf{C}(\mathbf{x} + \mathbf{y}) = \mathbf{B}[\mathbf{A}(\mathbf{x} + \mathbf{y})] = \mathbf{B}(\mathbf{Ax} + \mathbf{Ay}) = \mathbf{B}(\mathbf{Ax}) + \mathbf{B}(\mathbf{Ay}) = \mathbf{Cx} + \mathbf{Cy},$$

$$\mathbf{C}(\lambda\mathbf{x}) = \mathbf{B}[\mathbf{A}(\lambda\mathbf{x})] = \mathbf{B}(\lambda\mathbf{Ax}) = \lambda\mathbf{B}(\mathbf{Ax}) = \lambda\mathbf{Cx}.$$

The transformation

$$\mathbf{C} = \mathbf{BA}$$

is called the *product* of the transformations $\mathbf{A}$ and $\mathbf{B}$, where the factors are written from right to left in the order in which the corresponding transformations are carried out.

THEOREM 1. *Multiplication of linear transformations is associative, i.e.,*

$$\mathbf{C}(\mathbf{BA}) = (\mathbf{CB})\mathbf{A}.$$

Proof. Given any $\mathbf{x} \in L_3$, we have

$$[\mathbf{C}(\mathbf{BA})]\mathbf{x} = \mathbf{C}[(\mathbf{BA})\mathbf{x}] = \mathbf{C}[\mathbf{B}(\mathbf{Ax})] = (\mathbf{CB})(\mathbf{Ax}) = [(\mathbf{CB})\mathbf{A}]\mathbf{x}. \quad ▮$$

THEOREM 2. *The product of a linear transformation with the identity transformation* (*in either order*) *is the transformation itself, i.e.,*

$$\mathbf{AE} = \mathbf{EA} = \mathbf{A}.$$

Proof. We need merely note that

$$(\mathbf{AE})\mathbf{x} = \mathbf{A}(\mathbf{Ex}) = \mathbf{Ax} = \mathbf{E}(\mathbf{Ax}) = (\mathbf{EA})\mathbf{x}. \quad ▮$$

Remark. In other words, the identity transformation serves as the unit for operator multiplication.

THEOREM 3. *Multiplication of linear transformations is noncommutative, i.e., in general*

$$\mathbf{AB} \neq \mathbf{BA}.$$

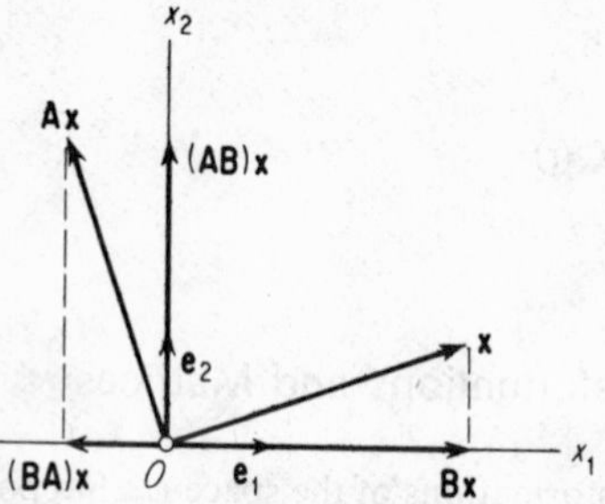

FIGURE 12

Proof. It is enough to give an example where $\mathbf{AB} \neq \mathbf{BA}$. Let **A** be rotation of the plane L_2 through 90° about the point O, and let **B** be projection of L_2 onto the x_1-axis. Then, given any vector $\mathbf{x} \in L_2$, Figure 12 shows that the vector $(\mathbf{BA})\mathbf{x}$ lies along the x_1-axis, while the vector $(\mathbf{AB})\mathbf{x}$ lies along the x_2-axis. It follows that

$$(\mathbf{AB})\mathbf{x} \neq (\mathbf{BA})\mathbf{x},$$

and hence $\mathbf{AB} \neq \mathbf{BA}$. ∎

Two transformations **A** and **B** are said to *commute* if $\mathbf{AB} = \mathbf{BA}$. For example according to Theorem 2, every transformation **A** commutes with the identity transformation **E**. As another example, let **A** be a transformation expanding the plane along the x_1-axis and **B** a transformation expanding the plane along the x_2-axis. Then **A** and **B** commute, since

$$\mathbf{Ax} = \lambda_1 x_1 \mathbf{e}_1 + x_2 \mathbf{e}_2,$$
$$\mathbf{Bx} = x_1 \mathbf{e}_1 + \lambda_2 x_2 \mathbf{e}_2,$$

and hence

$$(\mathbf{AB})\mathbf{x} = \lambda_1 x_1 \mathbf{e}_1 + \lambda_2 x_2 \mathbf{e}_2 = (\mathbf{BA})\mathbf{x}.$$

16.2. Suppose the linear transformations **A** and **B** have matrices A and B relative to some basis $\mathbf{e}_1, \mathbf{e}_2, \mathbf{e}_3$ in the space L_3, and suppose the product transformation $\mathbf{C} = \mathbf{BA}$ has the matrix C in the same basis. Then the matrix C is called the *product* of the matrices A and B, denoted by

$$C = BA.$$

As before, the factors are written from right to left in the order in which the corresponding transformations are carried out.

To express the elements of the matrix C in terms of those of the matrices A and B, suppose $A = (a_{ij})$, $B = (b_{ij})$, $C = (c_{ij})$. Then the transformation $\mathbf{y} = \mathbf{Ax}$ has the component form

$$y_j = a_{jk} x_k \tag{1}$$

(in the basis $\mathbf{e}_1, \mathbf{e}_2, \mathbf{e}_3$), while the transformation $\mathbf{z} = \mathbf{By}$ has the form

$$z_i = b_{ij} y_j. \tag{2}$$

Substituting (2) into (1), we get the component form of the transformation $\mathbf{z} = \mathbf{Cx}$:

$$z_i = b_{ij}a_{jk}x_k.$$

Since

$$z_i = c_{ik}x_k,$$

we find that the elements of the matrix C are just

$$c_{ik} = b_{ij}a_{jk}. \tag{3}$$

Thus *the quantities c_{ik} are the components of the second-order tensor obtained by contracting the tensors b_{ij} and a_{jk} with respect to the index j.*

Equation (3) can be written in more detail as

$$c_{ik} = b_{i1}a_{1k} + b_{i2}a_{2k} + b_{i3}a_{3k}.$$

But

$$A = \begin{pmatrix} a_{11} & a_{12} & a_{13} \\ a_{21} & a_{22} & a_{23} \\ a_{31} & a_{32} & a_{33} \end{pmatrix}, \quad B = \begin{pmatrix} b_{11} & b_{12} & b_{13} \\ b_{21} & b_{22} & b_{23} \\ b_{31} & b_{32} & b_{33} \end{pmatrix},$$

and hence *the element c_{ik} of the matrix C is obtained by multiplying the element of the ith row of the matrix B by the corresponding element of the jth column of the matrix A and then adding the resulting products.*

Remark 1. Multiplication of square matrices of any order can be defined in just the same way. For example, for second-order matrices we have

$$\begin{pmatrix} a_{11} & a_{12} \\ a_{21} & a_{22} \end{pmatrix}\begin{pmatrix} b_{11} & b_{12} \\ b_{21} & b_{22} \end{pmatrix} = \begin{pmatrix} a_{11}b_{11} + a_{12}b_{21} & a_{11}b_{12} + a_{12}b_{22} \\ a_{21}b_{11} + a_{22}b_{21} & a_{21}b_{12} + a_{22}b_{22} \end{pmatrix}.$$

Remark 2. All the basic results for multiplication of linear transformations carry over automatically to the case of multiplication of matrices, with the matrix $E = (\delta_{ij})$ of the identity transformation playing the role of multiplicative unit (this is why E is called the *unit matrix* in Example 1, p. 70). Like multiplication of linear transformations, multiplication of matrices is noncommutative. For example,

$$\begin{pmatrix} 1 & 2 \\ 0 & 1 \end{pmatrix}\begin{pmatrix} 3 & 0 \\ -1 & 1 \end{pmatrix} = \begin{pmatrix} 1\cdot 3 + 2\cdot -1 & 1\cdot 0 + 2\cdot 1 \\ 0\cdot 3 + 1\cdot -1 & 0\cdot 0 + 1\cdot 1 \end{pmatrix} = \begin{pmatrix} 1 & 2 \\ -1 & 1 \end{pmatrix},$$

$$\begin{pmatrix} 3 & 0 \\ -1 & 1 \end{pmatrix}\begin{pmatrix} 1 & 2 \\ 0 & 1 \end{pmatrix} = \begin{pmatrix} 3\cdot 1 + 0\cdot 0 & 3\cdot 2 + 0\cdot 1 \\ -1\cdot 1 + 1\cdot 0 & -1\cdot 2 + 1\cdot 1 \end{pmatrix} = \begin{pmatrix} 3 & 6 \\ -1 & -1 \end{pmatrix}.$$

Next we find the linear transformation adjoint to a product of linear transformations:

THEOREM. *If* $\mathbf{A}$ *and* $\mathbf{B}$ *are two linear transformations, then*

$$(\mathbf{AB})^* = \mathbf{B}^*\mathbf{A}^*. \tag{4}$$

Proof. Quite generally,

$$(\mathbf{x}, \mathbf{A}\mathbf{y}) = (\mathbf{y}, \mathbf{A}^*\mathbf{x}) = (\mathbf{A}^*\mathbf{x}, \mathbf{y})$$

for arbitrary vectors **x** and **y**. Hence

$$(\mathbf{x}, (\mathbf{A}\mathbf{B})\mathbf{y}) = (\mathbf{x}, \mathbf{A}(\mathbf{B}\mathbf{y})) = (\mathbf{A}^*\mathbf{x}, \mathbf{B}\mathbf{y}) = (\mathbf{B}^*(\mathbf{A}^*\mathbf{x}), \mathbf{y}) = ((\mathbf{B}^*\mathbf{A}^*)\mathbf{x}, \mathbf{y})$$

for all **x** and **y**, which implies (4). ∎

Remark. The matrix analogue of (4) is just

$$(AB)^* = B^*A^*,$$

where the asterisk now denotes the operation of *transposition.*

16.3. Next we prove the following key

THEOREM. *If A and B are two matrices with determinants $|A|$ and $|B|$, then*

$$|AB| = |A||B|,$$

where $|AB|$ is the determinant of the product matrix AB.

Proof. Let **A** and **B** be the linear transformations corresponding to A and B in an underlying orthonormal basis $\mathbf{e}_1, \mathbf{e}_2, \mathbf{e}_3$. Then the product matrix $C = AB$ corresponds to the product transformation $\mathbf{C} = \mathbf{AB}$. Let V_x be the oriented volume of the parallelepiped constructed on arbitrary vectors $\mathbf{x}_1, \mathbf{x}_2, \mathbf{x}_3 \in L_3$. Then **B** carries the vectors $\mathbf{x}_i$ into the vectors $\mathbf{y}_i = \mathbf{B}\mathbf{x}_i$ "spanning" a parallelepiped of volume

$$V_y = |B| V_x \tag{5}$$

(recall Sec. 14.3). By the same token, **A** carries the vectors $\mathbf{y}_i$ into the vectors $\mathbf{z}_i = \mathbf{A}\mathbf{y}_i$ spanning a parallelepiped of volume

$$V_z = |A| V_y. \tag{6}$$

On the other hand, $\mathbf{z}_i = \mathbf{C}\mathbf{x}_i$ and hence

$$V_z = |C| V_x,$$

where $|C|$ is the determinant of the matrix C. Substituting (5) into (6) and comparing the result with (7), we get

$$|C| = |AB| = |A||B|. \quad ∎$$

Remark 1. Applying the theorem twice, we find that

$$|AB| = |A||B| = |B||A| = |BA|,$$

i.e., the determinant of the product of two matrices does not depend on the order of the factors. In particular, if one of the transformations **A** and **B** is singular, as defined on p. 75, then so is their product (in either order).

Remark 2. The above theorem on multiplication of determinants can be proved purely algebraically, by using familiar properties of determinants.

For example, for two second-order determinants

$$A = \begin{pmatrix} a_{11} & a_{12} \\ a_{21} & a_{22} \end{pmatrix}, \qquad B = \begin{pmatrix} b_{11} & b_{12} \\ b_{21} & b_{22} \end{pmatrix},$$

we have

$$C = BA = \begin{pmatrix} b_{11}a_{11} + b_{12}a_{21} & b_{11}a_{12} + b_{12}a_{22} \\ b_{21}a_{11} + b_{22}a_{21} & b_{21}a_{12} + b_{22}a_{22} \end{pmatrix},$$

so that

$$\begin{aligned} |C| &= \begin{vmatrix} b_{11}a_{11} + b_{12}a_{21} & b_{11}a_{12} + b_{12}a_{22} \\ b_{21}a_{11} + b_{22}a_{21} & b_{21}a_{12} + b_{22}a_{22} \end{vmatrix} \\ &= \begin{vmatrix} b_{11}a_{11} & b_{11}a_{12} \\ b_{21}a_{11} & b_{21}a_{12} \end{vmatrix} + \begin{vmatrix} b_{11}a_{11} & b_{12}a_{22} \\ b_{21}a_{11} & b_{22}a_{22} \end{vmatrix} \\ &+ \begin{vmatrix} b_{12}a_{21} & b_{11}a_{12} \\ b_{22}a_{21} & b_{21}a_{12} \end{vmatrix} + \begin{vmatrix} b_{12}a_{21} & b_{12}a_{22} \\ b_{22}a_{21} & b_{22}a_{22} \end{vmatrix}. \end{aligned}$$

The first and last of the determinants on the right vanish since their columns are proportional, and hence

$$\begin{aligned} |C| &= a_{11}a_{22}\begin{vmatrix} b_{11} & b_{12} \\ b_{21} & b_{22} \end{vmatrix} + a_{21}a_{12}\begin{vmatrix} b_{12} & b_{11} \\ b_{22} & b_{21} \end{vmatrix} \\ &= \begin{vmatrix} a_{11} & a_{12} \\ a_{21} & a_{22} \end{vmatrix}\begin{vmatrix} b_{11} & b_{12} \\ b_{21} & b_{22} \end{vmatrix} = |B||A|. \end{aligned}$$

16.4. Using matrix multiplication, we now derive a new form of the transformation law for the elements of the matrix $A = (a_{ij})$ of a linear operator **A** under a change of basis. It will be recalled from Sec. 15.1 that a_{ij} is a second-order tensor. Hence, under the transformation

$$\mathbf{e}_{i'} = \gamma_{i'i}\mathbf{e}_i$$

from one orthonormal basis $\mathbf{e}_1, \mathbf{e}_2, \mathbf{e}_3$ to another orthonormal basis $\mathbf{e}_{1'}, \mathbf{e}_{2'}, \mathbf{e}_{3'}$, the quantity a_{ij} transforms according to the law

$$a_{i'j'} = \gamma_{i'i}\gamma_{j'j}a_{ij}$$

(see Sec. 9.3), where the $\gamma_{i'i}$ are the elements of the orthogonal matrix $\Gamma = (\gamma_{i'i})$ describing the basis transformation. For the orthogonal matrix Γ we have

$$\gamma_{i'i} = \gamma_{ii'}$$

(see Sec. 6.1), where the $\gamma_{ii'}$ are the elements of the matrix Γ^{-1} describing the inverse transformation from the new basis back to the old basis. It follows that

$$a_{i'j'} = \gamma_{i'i}a_{ij}\gamma_{jj'}. \tag{7}$$

But the right-hand side of (7) is just the result (in "element form") of multiplying the matrices Γ, A and Γ^{-1}. In fact, if A' denotes the matrix of the linear transformation **A** in the new basis $\mathbf{e}_{1'}, \mathbf{e}_{2'}, \mathbf{e}_{3'}$, then the matrix version of (4) is just

$$A' = \Gamma A \Gamma^{-1}. \tag{8}$$

This way of writing the transformation of the matrix of a linear operator in going over to a new basis is particularly convenient.

Remark 1. It follows from (8) that the determinant of the matrix of a linear transformation does not change in going over to a new basis. In fact, by (8) and the theorem on multiplication of determinants,

$$|A'| = |\Gamma||A||\Gamma^{-1}|.$$

But

$$|\Gamma| = |\Gamma^{-1}| = \pm 1$$

(see p. 24), and hence

$$|A'| = |A|. \tag{9}$$

Formula (9) shows that the determinant of a linear transformation is an invariant, and hence must have a well-defined geometric meaning. In fact, as we saw in Sec. 14.3, the determinant of a linear transformation is just the "magnification" of volumes under the transformation.

Remark 2. The matrix $\Gamma = (\gamma_{i'i})$ describing the transformation from the basis $\mathbf{e}_1, \mathbf{e}_2, \mathbf{e}_3$ to the new basis $\mathbf{e}_{1'}, \mathbf{e}_{2'}, \mathbf{e}_{3'}$ is not a tensor, since its indices i and i' pertain to different coordinate systems (in particular, Γ does not define a bilinear form on the space L_3).

PROBLEMS

1. Verify that the following formulas hold for linear transformations (and for matrices with boldface changed to lightface):

a) $\lambda(\mathbf{AB}) = (\lambda\mathbf{A})\mathbf{B}$;
b) $(\mathbf{A} + \mathbf{B})\mathbf{C} = \mathbf{AC} + \mathbf{BC}$;
c) $\mathbf{C}(\mathbf{A} + \mathbf{B}) = \mathbf{CA} + \mathbf{CB}$;
d) $\mathbf{A}^m\mathbf{A}^n = \mathbf{A}^{m+n}$;
e) $(\mathbf{A} + \mathbf{B})^2 = \mathbf{A}^2 + \mathbf{AB} + \mathbf{BA} + \mathbf{B}^2$;
f) $(\mathbf{A} + \mathbf{B})^3 = \mathbf{A}^3 + \mathbf{A}^2\mathbf{B} + \mathbf{ABA} + \mathbf{AB}^2 + \mathbf{BA}^2 + \mathbf{BAB} + \mathbf{B}^2\mathbf{A} + \mathbf{B}^3$;
g) $(\mathbf{A} + \mathbf{B})(\mathbf{A} - \mathbf{B}) = \mathbf{A}^2 + \mathbf{BA} - \mathbf{AB} - \mathbf{B}^2$.

What happens to the last three formulas if $\mathbf{AB} = \mathbf{BA}$?

2. Prove that the transformation **A** equal to the product of two expansions (compressions) of a rectangular coordinate system along the x_1 and x_2-axes with coefficients k and $1/k$, respectively, carries the family of hyperbolas $x_1x_2 = c$ into itself. Find the matrix of this transformation, and show that it does not change areas of figures.

3. Prove that the transformation **A** equal to the product of an expansion (compression) along the x_1-axis with coefficient a_2/a_1, a rotation through the angle α, and an expansion (compression) along the x_1-axis with coefficient a_1/a_2 (in that order) carries the ellipse

$$\frac{x_1^2}{a_1^2} + \frac{x_2^2}{a_2^2} = 1 \tag{10}$$

and ellipses homothetic to (10), into themselves. Find the matrix of the transformation, and show that it does not change areas of figures.

4. Prove that

a) $\begin{pmatrix} 1 & 1 \\ 0 & 1 \end{pmatrix}^n = \begin{pmatrix} 1 & n \\ 0 & 1 \end{pmatrix}$; b) $\begin{pmatrix} \lambda & 1 \\ 0 & \lambda \end{pmatrix}^n = \begin{pmatrix} \lambda^n & n\lambda^{n-1} \\ 0 & \lambda^n \end{pmatrix}$;

c) $\begin{pmatrix} \cos\theta & -\sin\theta \\ \sin\theta & \cos\theta \end{pmatrix}^n = \begin{pmatrix} \cos n\theta & -\sin n\theta \\ \sin n\theta & \cos n\theta \end{pmatrix}$.

5. Find A^n for the matrix

$$A = \begin{pmatrix} \lambda_1 & 0 & 0 \\ 0 & \lambda_2 & 0 \\ 0 & 0 & \lambda_3 \end{pmatrix}.$$

6. Prove that if two symmetric matrices commute, then their product is symmetric.

7. Prove that if A and B are antisymmetric matrices and if $AB = -BA$, then AB is antisymmetric.

8. Prove that

$$(\mathbf{Ax}, \mathbf{By}) = (\mathbf{x}, (\mathbf{A}^*\mathbf{B})\mathbf{y}) = (\mathbf{y}, \mathbf{B}^*\mathbf{Ax})$$

for arbitrary linear transformations **A**, **B** and arbitrary vectors **x**, **y**.

9. Prove that if **A** is a linear transformation, then **AA*** is a symmetric transformation.

10. Prove that the product of two orthogonal matrices (see Sec. 6.1) is itself an orthogonal matrix.

11. By the *trace* of a square matrix is meant the sum of the elements along its main diagonal (cf. p. 52). Given two square matrices A and B, prove that the trace of the product AB equals the trace of the product BA.

12. Prove that the rank of the product of an arbitrary (linear) transformation **A** and a nonsingular transformation **B** equals the rank of **A**.

13. Prove that the following formulas hold for arbitrary linear transformations **A** and **B** of a linear space L (or for the corresponding matrices, with boldface changed to lightface):

a) rank of $\mathbf{A} + \mathbf{B} \leq$ rank of $\mathbf{A}$ + rank of $\mathbf{B}$;

b) defect of $\mathbf{AB} \leq$ defect of $\mathbf{A}$ + defect of $\mathbf{B}$;

c) rank of $\mathbf{AB} \leq$ rank of **A**, rank of $\mathbf{AB} \leq$ rank of **B**

(see Prob. 16, p. 78).

14. Prove that if a (square) matrix A has the property that $AB = BA$ for every matrix B (of the same order), then $A = \lambda E$.

15. Prove that if a matrix A has the property that $AB = BA$ for every diagonal matrix B, then A is a diagonal matrix.

16. Find all matrices which commute with the following matrices:

$$\text{a)}\ \begin{pmatrix} 1 & 2 \\ 3 & 4 \end{pmatrix}; \qquad \text{b)}\ \begin{pmatrix} 3 & 1 & 0 \\ 0 & 3 & 1 \\ 0 & 0 & 3 \end{pmatrix}.$$

17. Find all second-order matrices A such that $A^2 = N$, where N is the null matrix.

18. A matrix A is said to be *involutory* if $A^2 = E$ and *idempotent* if $A^2 = A$. Find all involutory matrices of order two.

19. Prove that if a matrix has two of the three properties of being symmetric, orthogonal and involutory, then it has the third property.

20. Which of the following matrices are idempotent:

$$A_1 = \begin{pmatrix} 25 & -20 \\ 30 & -24 \end{pmatrix}, \qquad A_2 = \begin{pmatrix} 1 & 0 & 0 \\ 0 & 1 & 0 \\ 0 & 0 & 0 \end{pmatrix}?$$

21. Prove that if B is an idempotent matrix, then the matrix

$$A = 2B - E$$

is involutory, and conversely if A is involutory, then

$$B = \tfrac{1}{2}(A + E)$$

is idempotent.

22. Let $\mathbf{A}$ be the differentiation operator in the space of all polynomials $P(t)$ of degree not exceeding n. Prove that $\mathbf{A}^{n+1} = \mathbf{N}$, where $\mathbf{N}$ is the null transformation. Find the matrices of $\mathbf{A}^2, \mathbf{A}^3, \ldots$ in the basis $1, t, t^2, \ldots, t^n$. Find the null space, range, defect, and rank of $\mathbf{A}^2, \mathbf{A}^3, \ldots$ (cf. Prob. 19, p. 78).

23. Let $\mathbf{A}$ be the differentiation operator in the space of *all* polynomials $P(t)$ (of arbitrary degree), and let $\mathbf{B}$ be the operator of multiplication by the independent variable t:

$$\mathbf{A}[P(t)] = P'(t), \qquad \mathbf{B}[P(t)] = tP(t).$$

Prove that

a) $\mathbf{AB} - \mathbf{BA} = \mathbf{E}$;

b) $\mathbf{AB}^n - \mathbf{B}^n\mathbf{A} = n\mathbf{B}^{n-1}$.

Why can't the transformation $\mathbf{B}$ be considered in the space of all polynomials of degree not exceeding n?

17. Inverse Transformations and Matrices

17.1. Given a linear transformation $\mathbf{y} = \mathbf{Ax}$, the transformation $\mathbf{B}$ is called the *inverse* (*transformation*) of $\mathbf{A}$ if

$$\mathbf{By} = \mathbf{B}(\mathbf{Ax}) = \mathbf{x},$$

i.e., if **B** carries the vector **y** back into the original vector **x**. Thus the inverse transformation **B** is defined by the equation

$$\mathbf{BA} = \mathbf{E},$$

where **E** is the identity transformation. It is easy to see that the transformation **B** is itself linear (give the details).

Not every linear transformation **A** has an inverse. For example, let **A** be the transformation projecting the space L_3 onto some plane Π. Then the image **y** of every spatial vector **x** lies in Π, and a vector **y** not in Π has no inverse image **x**, while a vector **y** in Π has infinitely many inverse images! However, as we will see in a moment, every *nonsingular* transformation has an inverse.

The transformation inverse to the transformation **A** is denoted by $\mathbf{A}^{-1}$, so that

$$\mathbf{A}^{-1}\mathbf{A} = \mathbf{E}.$$

It is obvious that

$$(\mathbf{A}^{-1})^{-1} = \mathbf{A}, \quad \mathbf{A}\mathbf{A}^{-1} = \mathbf{E}.$$

Suppose the transformation **A** has an inverse, and let A be the matrix of the transformation **A** in some basis $\mathbf{e}_1, \mathbf{e}_2, \mathbf{e}_3$. Then the matrix of the transformation $\mathbf{A}^{-1}$ is called the *inverse* (*matrix*) of the matrix A and is denoted by A^{-1}. Since the matrix of a product transformation is the product of the matrices of the factors, we have

$$A^{-1}A = E, \quad AA^{-1} = E,$$

where E is the unit matrix. It follows from the theorem on multiplication of determinants (see Sec. 16.3) that

$$|A^{-1}||A| = 1,$$

i.e., the product of the determinants of a matrix and its inverse equals 1. Hence if a matrix A has an inverse, its determinant $|A|$ must be nonvanishing.

17.2. Next we prove the proposition mentioned above:

THEOREM. *If* **A** *is a nonsingular linear transformation, then* **A** *has a unique inverse* $\mathbf{A}^{-1}$.

Proof. Relative to some basis $\mathbf{e}_1, \mathbf{e}_2, \mathbf{e}_3$, the transformation $\mathbf{y} = \mathbf{Ax}$ takes the form

$$y_i = a_{ij}x_j, \tag{1}$$

where (a_{ij}) is the matrix of **A** and $\mathbf{x} = x_i\mathbf{e}_i$, $\mathbf{y} = y_i\mathbf{e}_i$. Finding the inverse transformation $\mathbf{A}^{-1}$ means finding the vector **x** for every given vector **y**, or equivalently, finding the components of **x** given those of **y**, i.e., solving the system (1) for the unknowns x_j given the numbers y_i. But, by Cramer's rule,† this system (consisting of three equations in the

† See e.g., R. A. Silverman, *op. cit.*, p. 610.

three unknowns x_1, x_2, x_3) has a unique solution if the determinant of the system is nonvanishing, i.e., if **A** is nonsingular. ∎

Next we find the matrix A^{-1} of the transformation $\mathbf{A}^{-1}$ inverse to **A**, assuming that $|A| \neq 0$. To this end, we write the system (1) in the more detailed form

$$\begin{aligned} a_{11}x_1 + a_{12}x_2 + a_{13}x_3 &= y_1, \\ a_{21}x_1 + a_{22}x_2 + a_{23}x_3 &= y_2, \\ a_{31}x_1 + a_{32}x_2 + a_{33}x_3 &= y_3. \end{aligned} \tag{1'}$$

Since the determinant of the system (1′) is nonsingular, we can apply Cramer's rule, obtaining, for example,

$$x_1 = \frac{1}{|A|}\begin{vmatrix} y_1 & a_{12} & a_{13} \\ y_2 & a_{22} & a_{23} \\ y_3 & a_{32} & a_{33} \end{vmatrix}. \tag{2}$$

Let A_{ij} denote the cofactor† of the element a_{ij} in the determinant $|A|$. Then (2) becomes

$$x_1 = \frac{A_{11}}{|A|}y_1 + \frac{A_{21}}{|A|}y_2 + \frac{A_{31}}{|A|}y_3.$$

Similarly, we have

$$x_2 = \frac{A_{12}}{|A|}y_1 + \frac{A_{22}}{|A|}y_2 + \frac{A_{32}}{|A|}y_3,$$

$$x_3 = \frac{A_{13}}{|A|}y_1 + \frac{A_{23}}{|A|}y_2 + \frac{A_{33}}{|A|}y_3.$$

The coefficients of y_j appearing in these expansions are the elements of the required inverse matrix A^{-1}. Thus, if $A^{-1} = (\tilde{a}_{ij})$, we have

$$\tilde{a}_{ij} = \frac{A_{ji}}{|A|},$$

i.e., *the element $\tilde{a}_{ij}$ of the inverse matrix A^{-1} equals the cofactor of the element a_{ji} of the original matrix A, divided by the determinant of A.*

Remark 1. Clearly, $\tilde{a}_{ij}$ is a second-order tensor, corresponding to the inverse transformation $\mathbf{A}^{-1}$. The tensor $\tilde{a}_{ij}$ is called the inverse of the tensor a_{ij} corresponding to the original transformation **A**.

Remark 2. The inverse A^{-1} of a second-order matrix

$$A = \begin{pmatrix} a_{11} & a_{12} \\ a_{21} & a_{22} \end{pmatrix}$$

is defined by the usual condition

$$A^{-1}A = AA^{-1} = E, \tag{3}$$

where E is now the unit matrix of order two. Here

† *Ibid.*, p. 602.

$$A_{11} = a_{22}, \qquad A_{12} = -a_{21}, \qquad A_{21} = -a_{12}, \qquad A_{22} = a_{11},$$

and hence

$$A^{-1} = \begin{pmatrix} \dfrac{a_{22}}{|A|} & -\dfrac{a_{12}}{|A|} \\[2ex] \dfrac{a_{21}}{|A|} & -\dfrac{a_{11}}{|A|} \end{pmatrix}.$$

For example, if

$$A = \begin{pmatrix} 2 & 1 \\ 3 & 2 \end{pmatrix},$$

then $|A| = 1$ and

$$A^{-1} = \begin{pmatrix} 2 & -1 \\ -3 & 2 \end{pmatrix}.$$

The validity of (3) follows at once by direct multiplication.

We now write some relations satisfied by the elements of the matrix $A = (a_{ij})$ and its inverse $A^{-1} = (\tilde{a}_{ij})$. Clearly $A^{-1}A = E$ implies

$$\tilde{a}_{ik}a_{kj} = \delta_{ij},$$

while $AA^{-1} = E$ implies

$$a_{ik}\tilde{a}_{kj} = \delta_{ij},$$

where, as usual, summation over k is understood on the left and δ_{ij} is the symmetric Kronecker symbol. It should also be noted that

$$(\mathbf{B}^{-1}\mathbf{A}^{-1})(\mathbf{AB}) = \mathbf{B}^{-1}(\mathbf{A}^{-1}\mathbf{A})\mathbf{B} = \mathbf{B}^{-1}\mathbf{E}\mathbf{B} = \mathbf{B}^{-1}\mathbf{B} = \mathbf{E},$$

and hence

$$(\mathbf{AB})^{-1} = \mathbf{B}^{-1}\mathbf{A}^{-1}$$

(there is an obvious analogue for matrices). Finally we note that, as implied by the notation, the matrix Γ^{-1}, the matrix of the transformation from the new basis $\mathbf{e}_{1'}, \mathbf{e}_{2'}, \mathbf{e}_{3'}$ to the old basis $\mathbf{e}_1, \mathbf{e}_2, \mathbf{e}_3$ (see pp. 23, 91), is just the inverse of the matrix Γ, the matrix of the transformation from the old basis to the new basis, so that in particular

$$\Gamma^{-1}\Gamma = \Gamma\Gamma^{-1} = E.$$

PROBLEMS

1. Find the inverse of each of the following matrices:

a) $\begin{pmatrix} 3 & 2 \\ 7 & 5 \end{pmatrix}$; b) $\begin{pmatrix} \cos\alpha & -\sin\alpha \\ \sin\alpha & \cos\alpha \end{pmatrix}$; c) $\begin{pmatrix} 1 & 1 & 1 \\ 0 & 1 & 1 \\ 0 & 0 & 1 \end{pmatrix}$;

d) $\begin{pmatrix} 1 & 0 & 0 \\ a & 1 & 0 \\ 0 & a & 1 \end{pmatrix}$; e) $\begin{pmatrix} 1 & 2 & -3 \\ -3 & 2 & 4 \\ 2 & -1 & 0 \end{pmatrix}$.

2. Solve the following "matrix equations":

a) $\begin{pmatrix} 5 & 3 \\ 3 & 2 \end{pmatrix}\begin{pmatrix} x_1 \\ x_2 \end{pmatrix} = \begin{pmatrix} 7 \\ 8 \end{pmatrix};$

b) $\begin{pmatrix} 1 & 2 & -3 \\ -3 & 2 & 4 \\ 2 & -1 & 0 \end{pmatrix}\begin{pmatrix} x_1 \\ x_2 \\ x_3 \end{pmatrix} = \begin{pmatrix} 1 \\ 3 \\ 4 \end{pmatrix};$

c) $AX = B$, where

$$A = \begin{pmatrix} 2 & -3 \\ 5 & 6 \end{pmatrix}, \qquad B = \begin{pmatrix} 4 & 1 \\ 2 & 7 \end{pmatrix}, \qquad X = \begin{pmatrix} x_1 & x_2 \\ x_3 & x_4 \end{pmatrix};$$

d) $XA = B$, where

$$A = \begin{pmatrix} 1 & 2 & -3 \\ -3 & 2 & 4 \\ 2 & -1 & 0 \end{pmatrix}, \qquad B = \begin{pmatrix} 1 & -3 & 0 \\ 10 & 2 & 7 \\ 10 & 7 & 8 \end{pmatrix}, \qquad X = \begin{pmatrix} x_1 & x_2 & x_3 \\ x_4 & x_5 & x_6 \\ x_7 & x_8 & x_9 \end{pmatrix}.$$

3. Prove that the following formulas hold for nonsingular linear transformations:

a) $(\mathbf{A}_1\mathbf{A}_2 \ldots \mathbf{A}_{k-1}\mathbf{A}_k)^{-1} = \mathbf{A}_k^{-1}\mathbf{A}_{k-1}^{-1} \ldots \mathbf{A}_2^{-1}\mathbf{A}_1^{-1}$;
b) $(\mathbf{A}^m)^{-1} = (\mathbf{A}^{-1})^m$;
c) $(\mathbf{A}^*)^{-1} = (\mathbf{A}^{-1})^*$.

What are the matrix analogues of these formulas?

4. Prove that the following four formulas are equivalent for nonsingular matrices:

$$AB = BA, \qquad AB^{-1} = B^{-1}A, \qquad A^{-1}B = BA^{-1}, \qquad A^{-1}B^{-1} = B^{-1}A^{-1}.$$

5. Prove that

a) The inverse of a nonsingular symmetric matrix is symmetric;
b) The inverse of a nonsingular antisymmetric matrix is antisymmetric;
c) The inverse of an orthogonal matrix is orthogonal.

6. Prove that the inverse of a nonsingular "triangular matrix"

$$A = \begin{pmatrix} a_{11} & a_{12} & a_{13} \\ 0 & a_{22} & a_{23} \\ 0 & 0 & a_{33} \end{pmatrix}$$

is a matrix of the same type.

18. The Group of Linear Transformations and Its Subgroups

18.1. The set of all nonsingular linear transformations of the three-dimensional space L_3, equipped with the operation of multiplication of transformations, has the following four key properties:

a) The set is *closed under multiplication*, i.e., if **A** and **B** are nonsingular transformations, then so is their product $\mathbf{C} = \mathbf{AB}$;

b) The operation of multiplication of transformations is *associative*, i.e., $\mathbf{A}(\mathbf{BC}) = (\mathbf{AB})\mathbf{C}$;
c) The set contains an *identity transformation* $\mathbf{E}$ such that $\mathbf{AE} = \mathbf{EA} = \mathbf{A}$ for every $\mathbf{A}$;
d) Every nonsingular transformation $\mathbf{A}$ has a unique *inverse transformation* $\mathbf{A}^{-1}$ such that $\mathbf{A}^{-1}\mathbf{A} = \mathbf{AA}^{-1} = \mathbf{E}$.

The set of nonsingular linear transformations is not the only set with these properties. For example, the set of all positive rational numbers equipped with the ordinary operation of multiplication has all four properties,† and the same is true of the set of all nonzero complex numbers equipped with multiplication. There are many other examples of sets of the same type. This leads to the following

DEFINITION. *Any set of elements equipped with a multiplicative operation satisfying properties* a)–d) *is called a* ***group.***

Thus the set of all nonsingular linear transformations of the space L_3 is a group, denoted by GL_3 and called the *full linear group of dimension three.*

Remark 1. Relative to any basis in L_3 there is a one-to-one correspondence between nonsingular linear transformations of L_3 and square matrices of order three with nonvanishing determinants. Hence the set of all such matrices is essentially identical with the set of all nonsingular linear transformations, and will again be called the full linear group of dimension three, denoted by GL_3.

Remark 2. In just the same way, the set of all nonsingular linear transformations of the two-dimensional plane L_2 is a group, denoted by GL_2 and called the *full linear group of dimension two.* The set of all square matrices of order two with nonvanishing determinants is essentially identical with this group. More generally, the set of all nonsingular linear transformations of the n-dimensional space L_n (or equivalently, the set of all square matrices of order n with nonvanishing determinants) is a group, denoted by GL_n and called the *full linear group of dimension n.*

18.2. As we now show, certain subsets of the full linear group GL_3, and not just the whole group GL_3, also form groups. In other words, there are subsets of GL_3 which are closed under multiplication and which, whenever they contain a transformation $\mathbf{A}$, also contain the inverse transformation $\mathbf{A}^{-1}$. Note that properties b) and c) are automatically satisfied in any such subset of GL_3. In fact, the associative property, being valid in the whole set, is obviously valid in any subset, while the identity transformation belongs to the subset, since the latter, by hypothesis, contains the inverse $\mathbf{A}^{-1}$ of any transformation $\mathbf{A}$ in the subset and hence also contains the product

† With obvious changes in terminology, e.g., "number" for "transformation," "positive" for "nonsingular," etc.

$\mathbf{AA}^{-1} = \mathbf{E}$. Subsets of GL_3 of this type are called *subgroups* of GL_3 (subgroups of an arbitrary group are defined in just the same way).

Example 1. Consider the subset of GL_3 consisting of all linear transformations which do not change the orientation of noncoplanar triples of vectors. The matrix A of any such transformation $\mathbf{A}$ has a positive determinant $|A| > 0$ (see Sec. 14.3). Clearly, the product of any two such transformations is a transformation of the same type, and the same is true of the inverse of any such transformation. Hence the set of all transformations which do not change the orientation of noncoplanar triples of vectors is a subgroup of the group GL_3. Moreover, the set of all third-order matrices with positive determinants is essentially identical with this subgroup. Note that the set of all third-order matrices with *negative* determinants does not form a group, since the product of two matrices with negative determinants is a matrix with a positive determinant.

Example 2. Suppose the transformation $\mathbf{A}$ with matrix A does not change the absolute value of the volume of the parallelepiped constructed on an arbitrary triple of vectors. Then the absolute value of the determinant A equals unity, i.e., $|A| = \pm 1$. The set of all transformations of this type (and of their matrices as well) obviously forms a group, called the *unimodular group*. In turn, the set of all transformations which preserve both the volume and the orientation of triples of vectors forms a subgroup of the unimodular group. For such transformations, we have $|A| = 1$.

Example 3. Consider the set of all rotations of the plane L_2 about the origin of coordinates. This set is a group, since the product of any two rotations is obviously a rotation, and the same is true of the transformation inverse to any rotation. To verify this by a formal calculation, let

$$A = \begin{pmatrix} \cos\alpha & -\sin\alpha \\ \sin\alpha & \cos\alpha \end{pmatrix}, \qquad B = \begin{pmatrix} \cos\beta & -\sin\beta \\ \sin\beta & \cos\beta \end{pmatrix}$$

be the matrices corresponding to rotations through the angles α and β, respectively (recall Example 6, p. 72). Then the matrix

$$\begin{aligned} AB &= \begin{pmatrix} \cos\alpha\cos\beta - \sin\alpha\sin\beta & -\cos\alpha\sin\beta - \sin\alpha\cos\beta \\ \sin\alpha\cos\beta + \cos\alpha\sin\beta & -\sin\alpha\sin\beta + \cos\alpha\cos\beta \end{pmatrix} \\ &= \begin{pmatrix} \cos(\alpha+\beta) & -\sin(\alpha+\beta) \\ \sin(\alpha+\beta) & \cos(\alpha+\beta) \end{pmatrix} \end{aligned}$$

corresponds to a rotation through the angle $\alpha + \beta$, while the matrix

$$A^{-1} = \begin{pmatrix} \cos\alpha & \sin\alpha \\ -\sin\alpha & \cos\alpha \end{pmatrix},$$

corresponding to rotation through the angle $-\alpha$, clearly satisfies the relation $A^{-1}A = AA^{-1} = E$ (verify this).

18.3. Next we consider another important subgroup of the full linear group, namely the subgroup of orthogonal transformations. A linear transformation **A** is said to be *orthogonal* if it preserves scalar products, i.e., if

$$(\mathbf{Ax}, \mathbf{Ay}) = (\mathbf{x}, \mathbf{y})$$

for arbitrary vectors **x** and **y**.

THEOREM 1. *Every orthogonal transformation* **A** *preserves lengths of vectors and angles between vectors.*

Proof. Let $\mathbf{u} = \mathbf{Ax}$, $\mathbf{v} = \mathbf{Ay}$, and let φ be the angle between the vectors **x** and **y**, while ψ is the angle between the vectors **u** and **v**. Then

$$(\mathbf{u}, \mathbf{u}) = (\mathbf{Ax}, \mathbf{Ax}) = (\mathbf{x}, \mathbf{x}),$$

so that $|\mathbf{u}|^2 = |\mathbf{x}|^2$, or equivalently $|\mathbf{u}| = |\mathbf{x}|$. Similarly $|\mathbf{v}| = |\mathbf{y}|$, and hence

$$\cos\varphi = \frac{(\mathbf{x}, \mathbf{y})}{|\mathbf{x}||\mathbf{y}|} = \frac{(\mathbf{u}, \mathbf{v})}{|\mathbf{u}||\mathbf{v}|} = \cos\psi$$

(see Sec. 4), since

$$(\mathbf{u}, \mathbf{v}) = (\mathbf{Ax}, \mathbf{Ay}) = (\mathbf{x}, \mathbf{y})$$

by the orthogonality of **A**. It follows that $\varphi = \psi$, since the angle between vectors varies only from 0 to π. ∎

THEOREM 2. *If a linear transformation* **A** *preserves lengths of vectors, then* **A** *is orthogonal.*

Proof. Let $\mathbf{u} = \mathbf{Ax}$, $\mathbf{v} = \mathbf{Ay}$, so that $\mathbf{u} + \mathbf{v} = \mathbf{A}(\mathbf{x} + \mathbf{y})$. Then

$$|\mathbf{u} + \mathbf{v}| = |\mathbf{x} + \mathbf{y}|,$$

by hypothesis, and hence

$$(\mathbf{u} + \mathbf{v}, \mathbf{u} + \mathbf{v}) = (\mathbf{x} + \mathbf{y}, \mathbf{x} + \mathbf{y}).$$

It follows that

$$(\mathbf{u}, \mathbf{u}) + 2(\mathbf{u}, \mathbf{v}) + (\mathbf{v}, \mathbf{v}) = (\mathbf{x}, \mathbf{x}) + 2(\mathbf{x}, \mathbf{y}) + (\mathbf{y}, \mathbf{y}).$$

But $|\mathbf{u}| = |\mathbf{x}|, |\mathbf{v}| = |\mathbf{y}|$, or equivalently $(\mathbf{u}, \mathbf{u}) = (\mathbf{x}, \mathbf{x})$, $(\mathbf{v}, \mathbf{v}) = (\mathbf{y}, \mathbf{y})$, and hence

$$(\mathbf{u}, \mathbf{v}) = (\mathbf{Ax}, \mathbf{Ay}) = (\mathbf{x}, \mathbf{y}),$$

i.e., **A** is orthogonal. ∎

THEOREM 3. *A linear transformation* **A** *is orthogonal if and only if*

$$\mathbf{A}^*\mathbf{A} = \mathbf{E}. \tag{1}$$

Proof. If $\mathbf{u} = \mathbf{Ax}$, $\mathbf{v} = \mathbf{Ay}$, then

$$(\mathbf{u}, \mathbf{v}) = (\mathbf{Ax}, \mathbf{Ay}) = (\mathbf{x}, (\mathbf{A}^*\mathbf{A})\mathbf{y})$$

(see Sec. 15.3 and Prob. 8, p. 93). If (1) holds, then $(\mathbf{u}, \mathbf{v}) = (\mathbf{x}, \mathbf{y})$ and **A** is orthogonal. Conversely, if **A** is orthogonal, then $(\mathbf{u}, \mathbf{v}) = (\mathbf{x}, \mathbf{y})$ and hence (1) holds. ∎

Remark. Formula (1), satisfied by an orthogonal transformation **A**, can also be written in the form

$$\mathbf{A}^* = \mathbf{A}^{-1}. \tag{2}$$

THEOREM 4. *The set of all orthogonal transformations of the space L_3 forms a subgroup O_3 of the full linear group GL_3.*

Proof. It is enough to show that if the transformations **A** and **B** are orthogonal, so that

$$\mathbf{A}^* = \mathbf{A}^{-1}, \quad \mathbf{B}^* = \mathbf{B}^{-1},$$

then so are the product transformation $\mathbf{C} = \mathbf{AB}$ and the inverse transformation $\mathbf{A}^{-1}$. But

$$\mathbf{C}^* = (\mathbf{AB})^* = \mathbf{B}^*\mathbf{A}^* = \mathbf{B}^{-1}\mathbf{A}^{-1} = (\mathbf{AB})^{-1} = \mathbf{C}^{-1},$$

which implies the orthogonality of **C**, while

$$(\mathbf{A}^{-1})^* = (\mathbf{A}^*)^* = \mathbf{A} = (\mathbf{A}^{-1})^{-1},$$

which implies the orthogonality of $\mathbf{A}^{-1}$. ∎

Remark. The group character of the set of orthogonal transformations can be proved purely geometrically. In fact, if the transformations **A** and **B** do not change lengths of vectors and angles between them, then the same is clearly true of the product **AB** and the inverse $\mathbf{A}^{-1}$.

We now consider the matrices of orthogonal transformations, called *orthogonal matrices.* Such matrices have already been encountered in Sec. 6.1 in our treatment of transformations from one orthonormal basis to another. It follows from (1) that the matrix $A = (a_{ij})$ of an orthogonal transformation **A** satisfies the condition

$$A^*A = E \tag{3}$$

and the equivalent condition

$$AA^* = E. \tag{3'}$$

Let a^*_{ij} denote the elements of the matrix A^*, so that $a^*_{ij} = a_{ji}$ (cf. p. 80). Then (3) and (3′) become

$$\begin{aligned} a^*_{ik}a_{kj} &= a_{ki}a_{kj} = \delta_{ij}, \\ a_{ik}a^*_{kj} &= a_{ik}a_{jk} = \delta_{ij} \end{aligned} \tag{4}$$

in component form. The relations (4) show that *the sum of the products of the elements of any row (or column) with the corresponding elements of any other row (or column) equals zero, while the sum of the squares of the elements of any row (or column) equals unity.* Note that the relations (4) differ only in notation from the relations (7), p. 23.

Remark 1. It has already been proved geometrically on p. 24 that the determinant of an orthogonal matrix equals ± 1. We now give a simple algebraic proof of this fact. It follows from the formula $A^*A = E$ and the theorem on multiplication of determinants that

$$|A^*A| = |A^*||A| = |A|^2 = 1,$$

since $|A^*| = |A|$ and $|E| = 1$. But then obviously $|A| = \pm 1$.

Remark 2. Orthogonal transformations whose matrices have determinant $+1$ preserve the orientation of triples of vectors and are said to be *proper.* As is easily verified, the set of all proper orthogonal transformations is itself a group, namely a subgroup (denoted by O_3^+) of the group O_3 of all orthogonal transformations. Orthogonal transformations whose matrices have determinant -1 change the orientation of triples of vectors and are said to be *improper.*† The set of all improper orthogonal transformations is clearly not a group (why not?). Every reflection of the space L_3 in a plane Π through the origin O is an improper orthogonal transformation. In fact, reflection in Π does not change lengths of vectors or angles between them, but it does reverse the orientation of triples of vectors. It can easily be shown that every improper orthogonal transformation is the product of a proper orthogonal transformation and a reflection in some plane (cf. Sec. 20, Probs. 6 and 7).

18.4. The subgroups of the full linear group considered so far all contain infinitely many elements, like the full linear group itself. But there also exist *finite subgroups* of the full linear group, i.e., subgroups containing only a finite number of elements. Of particular interest are certain subgroups of the orthogonal group called *symmetry groups,* which are of great importance in crystallography and other branches of physics. We now give some examples of symmetry groups in the plane and in space.

Example 1. Let $\mathbf{E}$ be the identity transformation, and let $\mathbf{A} = -\mathbf{E}$ be the transformation which reflects all vectors in the origin O. Then clearly

$$\mathbf{EE} = \mathbf{E}, \qquad \mathbf{EA} = \mathbf{AE} = \mathbf{A}, \qquad \mathbf{AA} = \mathbf{E}, \qquad \mathbf{E}^{-1} = \mathbf{E}, \qquad \mathbf{A}^{-1} = \mathbf{A}.$$

Hence the set of transformations consisting of the two elements $\mathbf{E}$ and $\mathbf{A}$ forms a group, since it is closed under multiplication and the operation of inversion. The "multiplication table" for the elements of this group can be written in the form‡

	E	A
E	E	A
A	A	E

† Similarly, an orthogonal transformation in the plane L_2 is said to be *proper* if its matrix has determinant $+1$ and *improper* if its matrix has determinant -1.

‡ In any such multiplication table, the first factor in a product appears on the left and the second factor appears on top.

Every figure with the point O as a center of symmetry is carried into itself by the transformations of the group. Note that the product of two elements of the group does not depend on the order of the factors; a group of this type is said to be *commutative*.

Example 2. Let $\mathbf{E}$ be the identity transformation in the plane, and let $\mathbf{A}$ be a rotation of the plane through the angle $2\pi/n$ (n a positive integer). Then the transformation $\mathbf{A}^k$ is a rotation through the angle $2\pi k/n$, so that in particular $\mathbf{A}^n = \mathbf{E}$. The transformations $\mathbf{E}, \mathbf{A}, \mathbf{A}^2, \ldots, \mathbf{A}^{n-1}$ form a group, since

$$\mathbf{A}^k\mathbf{A}^l = \mathbf{A}^{k+l} = \mathbf{A}^m,$$

where m is the remainder after dividing $k + l$ by n. This group, called the *cyclic group of order n*, is also commutative.

Example 3. Given three perpendicular axes a, b and c going through the origin O of the space L_3, let $\mathbf{A}$ be a rotation through the angle π about a, $\mathbf{B}$ a rotation through π about b, and $\mathbf{C}$ a rotation through π about c. The four transformations $\mathbf{E}, \mathbf{A}, \mathbf{B}$ and $\mathbf{C}$ then form a group with the following multiplication table:

	E	**A**	**B**	**C**
E	**E**	**A**	**B**	**C**
A	**A**	**E**	**C**	**B**
B	**B**	**C**	**E**	**A**
C	**C**	**B**	**A**	**E**

Since the product of any two elements of the group is independent of the order of the factors, the group is commutative. The transformations of this group carry any figure with a, b and c as axes of symmetry into itself.

Example 4. Given two perpendicular axes a and b through the origin O of the space L_3, let $\mathbf{A}$ be a rotation through the angle $2\pi/3$ about a, while $\mathbf{B}$ is a rotation through π about b. Then the six transformations

$$\mathbf{E}, \quad \mathbf{A}, \quad \mathbf{A}^2, \quad \mathbf{B}, \quad \mathbf{AB}, \quad \mathbf{A}^2\mathbf{B} \tag{5}$$

form a group. In fact, the transformation $\mathbf{B}_1 = \mathbf{AB}$ is a rotation through π about the axis b_1 into which the transformation $\mathbf{A}^2$ carries the axis b, and similarly, the transformation $\mathbf{B}_2 = \mathbf{A}^2\mathbf{B}$ is a rotation through π about the axis b_2 into which the transformation $\mathbf{A}$ carries the axis b. It follows that

$$\mathbf{B}^2 = \mathbf{B}_1^2 = \mathbf{B}_2^2 = \mathbf{E}.$$

Moreover, using the fact that $\mathbf{A}^3 = \mathbf{E}$, we have

$$\mathbf{BA} = (\mathbf{BA})^{-1} = \mathbf{A}^{-1}\mathbf{B}^{-1} = \mathbf{A}^2\mathbf{B} = \mathbf{B}_2,$$
$$\mathbf{BA}^2 = (\mathbf{BA}^2)^{-1} = \mathbf{A}^{-2}\mathbf{B}^{-1} = \mathbf{AB} = \mathbf{B}_1.$$

The multiplication table for the transformations (5) now takes the form

	E	A	A²	B	B₁	B₂
E	E	A	A^2	B	B_1	B_2
A	A	A^2	E	B_1	B_2	B
A^2	A^2	E	A	B_2	B	B_1
B	B	B_2	B_1	E	A^2	A
B_1	B_1	B	B_2	A	E	A^2
B_2	B_2	B_1	B	A^2	A	E

from which it is apparent that the transformations form a group as asserted. Since the multiplication table is not symmetric, the group is not commutative.

PROBLEMS

1. Determine which of the following sets of transformations of the plane L_2 are groups:

a) The set of all rotations about a point O;
b) The set of all reflections in all possible straight lines through O;
c) The set of all homothetic transformations with center at O and all possible expansion coefficients;
d) The set of all homothetic transformations with center at O, together with all rotations about O;
e) Reflection in a line through O, together with the identity transformation;
f) The set of rotations about O through the angles 120°, 240° and 360°;
g) The set of rotations about O through the angles 90°, 180°, 270° and 360°, together with reflection in two given perpendicular lines intersecting at O.

2. List all orthogonal transformations of the plane L_2 carrying each of the following figures into itself:

a) A rhombus;
b) A square;
c) An equilateral triangle;
d) A regular hexagon.

3. Determine which of the following sets of numbers are groups under the indicated operations:

a) The set of integers under addition;
b) The set of rational numbers under addition;
c) The set of complex numbers under addition;
d) The set of nonnegative integers under addition;
e) The set of even numbers under addition;
f) The set of numbers of the form 2^k (k an integer) under multiplication;
g) The set of nonzero rational numbers under multiplication;
h) The set of nonzero real numbers under multiplication;

i) The set of nonzero complex numbers under multiplication;
j) The set of integral multiples of a given positive integer n under addition.

4. Determine which of the following sets are groups:
a) The set of matrices of order three with real elements under addition;
b) The set of nonsingular matrices of order three with real elements under multiplication;
c) The set of matrices of order three with integral elements under multiplication;
d) The set of matrices of order three with integral elements and determinants equal to ± 1 under multiplication;
e) The set of polynomials of degree not exceeding n in a variable x (excluding zero) under addition;
f) The set of polynomials of degree n under addition;
g) The set of polynomials of arbitrary degree (excluding zero) under addition.

6. Prove that a matrix A is orthogonal if and only if its determinant equals ± 1 and every element equals its own cofactor, taken with the plus sign if $|A| = 1$ and the minus sign if $|A| = -1$.

7. Under what conditions is a diagonal matrix orthogonal?

8. Find the matrices of the transformations considered in Examples 1–3 of Sec. 18.4, and verify by direct calculation that each set of matrices forms a group.

4

FURTHER TOPICS

19. Eigenvectors and Eigenvalues

19.1. Given a linear transformation $\mathbf{u} = \mathbf{Ax}$, a nonzero vector $\mathbf{x}$ is called an *eigenvector* of $\mathbf{A}$ if

$$\mathbf{Ax} = \lambda \mathbf{x}, \tag{1}$$

where λ is a real number. The number λ is then called an *eigenvalue* of the transformation $\mathbf{A}$, corresponding to the eigenvector $\mathbf{x}$. According to this definition, the transformation $\mathbf{A}$ carries an eigenvector into a collinear vector, with the corresponding eigenvalue equal to the ratio of the two collinear vectors (the "expansion coefficient" of the eigenvector under the transformation $\mathbf{A}$).

Obviously, if $\mathbf{x}$ is an eigenvector of $\mathbf{A}$ with eigenvalue λ, then any vector $\mathbf{x}' = \alpha\mathbf{x}$ collinear with $\mathbf{x}$ (α a nonzero real number) is also an eigenvector of $\mathbf{A}$ with eigenvalue λ. In fact, by the linearity of $\mathbf{A}$, we have

$$\mathbf{Ax}' = \mathbf{A}(\alpha\mathbf{x}) = \alpha(\mathbf{Ax}) = \alpha(\lambda\mathbf{x}) = \lambda(\alpha\mathbf{x}) = \lambda\mathbf{x}'.$$

Remark 1. Equation (1) can be written in the equivalent form

$$(\mathbf{A} - \lambda\mathbf{E})\mathbf{x} = \mathbf{0}, \tag{1'}$$

where $\mathbf{E}$ is the identity transformation.

Remark 2. Everything just said applies equally well to the plane L_2, to three-dimensional space L_3, or, more generally, to an arbitrary linear space L.

We now examine some of the linear transformations considered in Secs. 13 and 14 from the standpoint of eigenvectors and eigenvalues.

Example 1. For the homothetic transformation $\mathbf{Ax} = \lambda\mathbf{x}$ of the space L_3 (or L_2), every nonzero vector $\mathbf{x}$ is an eigenvector with eigenvalue λ. The same is obviously true of the identity transformation $\mathbf{E}$ ($\lambda = 1$) and of the operation of reflection in the origin ($\lambda = -1$).†

Example 2. For the transformation $\mathbf{A}$ carrying the vector

$$\mathbf{x} = x_1\mathbf{e}_1 + x_2\mathbf{e}_2 \in L_2$$

into the vector

$$\mathbf{u} = \mathbf{Ax} = x_1\mathbf{e}_1 + \lambda x_2\mathbf{e}_2$$

(see Example 5, p. 66), every vector lying on the x_1 or x_2-axes (all vectors have their initial points at the origin) is an eigenvector, with eigenvalue 1 for vectors on the x_1-axis and eigenvalue λ for vectors on the x_2-axis. In particular, the same vectors are eigenvectors for the operation of projection of the plane L_2 onto the x_1-axis ($\lambda = 0$), with the vectors on the x_2-axis all going into the zero vector $\mathbf{0}$ (which is collinear with every vector!).

Example 3. Let $\mathbf{A}$ be the rotation of the plane L_2 through an angle α different from 0° or 180°. Then $\mathbf{A}$ obviously has no real eigenvectors (however, see Example 1, p. 112). If, on the other hand, $\alpha = 0°$ or $\alpha = 180°$, we get the identity transformation or the operation of reflection in the origin, for which every vector is an eigenvector. By contrast, every rotation in the space L_3 has a unique real eigenvector, whose direction is that of the axis of rotation (cf. Prob. 2c, p. 115).

Example 4. For the shift

$$\mathbf{u} = \mathbf{Ax} = (x_1 + kx_2)\mathbf{e}_1 + x_2\mathbf{e}_2$$

of the plane L_2 in the direction of the vector $\mathbf{e}_1$ (see Example 8, p. 66), every vector lying on the x_1-axis is clearly an eigenvector with eigenvalue 1.

Example 5. The eigenvectors of the transformation

$$\mathbf{u} = \mathbf{Ax} = \lambda_1 x_1\mathbf{e}_1 + \lambda_2 x_2\mathbf{e}_2 + \lambda_3 x_3\mathbf{e}_3$$

of the space L_3, consisting of three simultaneous expansions along perpendicular axes $\mathbf{e}_1$, $\mathbf{e}_2$ and $\mathbf{e}_3$, are just the vectors lying along these axes, since

$$\mathbf{Ae}_i = \lambda_i\mathbf{e}_i \qquad \text{(no summation over } i\text{)},$$

and the corresponding eigenvalues are λ_1, λ_2 and λ_3. Similarly, the eigenvectors of the transformation

$$\mathbf{u} = \mathbf{Ax} = \lambda_1 x_1\mathbf{e}_1 + \lambda_2 x_2\mathbf{e}_2$$

in the plane L_2 are the vectors lying on the $\mathbf{e}_1$ and $\mathbf{e}_2$-axes, with eigenvalues λ_1 and λ_2.

19.2. Next we consider the problem of finding the eigenvectors and eigenvalues of a given linear transformation $\mathbf{A}$ of the space L_3. As we know

† See Examples 1 and 2, p. 65 and Prob. 2a, p. 67.

from Sec. 14.1, in any given orthonormal basis $\mathbf{e}_1, \mathbf{e}_2, \mathbf{e}_3$ the transformation $\mathbf{A}$ is associated with a matrix

$$A = \begin{pmatrix} a_{11} & a_{12} & a_{13} \\ a_{21} & a_{22} & a_{23} \\ a_{31} & a_{32} & a_{33} \end{pmatrix},$$

namely, the matrix of the transformation (relative to $\mathbf{e}_1, \mathbf{e}_2, \mathbf{e}_3$). Suppose $\mathbf{x} = x_i\mathbf{e}_i$ is an eigenvector of $\mathbf{A}$ with eigenvalue λ. Then, writing (1) in component form, we get

$$\begin{aligned} a_{11}x_1 + a_{12}x_2 + a_{13}x_3 &= \lambda x_1, \\ a_{21}x_1 + a_{22}x_2 + a_{23}x_3 &= \lambda x_2, \\ a_{31}x_1 + a_{32}x_2 + a_{33}x_3 &= \lambda x_3, \end{aligned} \tag{2}$$

or, more concisely,

$$a_{ij}x_j = \lambda x_i.$$

We can also write (2) as

$$\begin{aligned} (a_{11} - \lambda)x_1 + a_{12}x_2 + a_{13}x_3 &= 0, \\ a_{21}x_1 + (a_{22} - \lambda)x_2 + a_{23}x_3 &= 0, \\ a_{31}x_1 + a_{32}x_2 + (a_{33} - \lambda)x_3 &= 0, \end{aligned} \tag{3}$$

or, more concisely,

$$(a_{ij} - \lambda\delta_{ij})x_j = 0.$$

The system (3) is a system of three homogeneous linear equations in the three unknowns x_1, x_2 and x_3. Since, by hypothesis, (3) has a nontrivial solution, representing the components of the nonzero vector $\mathbf{x}$, the determinant of (3) must vanish,† i.e., we must have

$$\begin{vmatrix} a_{11} - \lambda & a_{12} & a_{13} \\ a_{21} & a_{22} - \lambda & a_{23} \\ a_{31} & a_{32} & a_{33} - \lambda \end{vmatrix} = 0, \tag{4}$$

or, briefly,

$$|A - \lambda E| = 0,$$

where E is the unit matrix.

This shows that every eigenvalue of the linear transformation $\mathbf{A}$ satisfies equation (4). Conversely, let λ_0 be a real root of equation (4). Then, substituting λ_0 for λ in (3), we get a system with a nontrivial solution x_1^0, x_2^0, x_3^0, since the determinant of the system vanishes.‡ Clearly, (3) holds for the vector $\mathbf{x}_0$ with components x_1^0, x_2^0, x_3^0, and hence

$$\mathbf{A}\mathbf{x}_0 = \lambda_0\mathbf{x}_0$$

† Otherwise (3) would only have the trivial solution $x_1 = x_2 = x_3 = 0$, by Cramer's rule.

‡ See e.g., R. A. Silverman, *op. cit.*, Prob. 11, p. 615, and its solution, p. 984.

for this vector and the number λ_0, i.e., $\mathbf{x}_0$ is an eigenvector of **A** corresponding to the eigenvalue λ_0.

Thus to find the eigenvectors of the transformation **A**, we must first solve equation (4). Each real root of (4) gives an eigenvalue of **A**, and the components of the eigenvector corresponding to this eigenvalue can then be determined from the system (3). Equation (4) is called the *characteristic* (or *secular*) *equation* of the transformation **A**.

Remark. So far we have only considered real scalars and vectors with real components. More generally, we might consider *complex* linear spaces, allowing both the scalars and the components of vectors to be complex numbers. In this context, we then allow complex eigenvalues and eigenvectors with complex components. Suppose the matrix of the transformation **A** has real elements, so that the characteristic equation (4) of **A** is an equation of degree three with real coefficients. Then, by elementary algebra, equation (4) either has three real roots, or else it has one real root and a pair of conjugate complex roots. Such a pair of conjugate complex roots will then correspond to a pair of conjugate complex eigenvectors of the transformation **A** (show this).

19.3. Suppose we expand the determinant appearing in the left-hand side of the characteristic equation (4). Then (4) takes the form

$$\lambda^3 - I_1\lambda^2 + I_2\lambda - I_3 = 0, \tag{5}$$

where

$$I_1 = a_{11} + a_{22} + a_{33},$$

$$I_2 = \begin{vmatrix} a_{11} & a_{12} \\ a_{21} & a_{22} \end{vmatrix} + \begin{vmatrix} a_{11} & a_{13} \\ a_{31} & a_{33} \end{vmatrix} + \begin{vmatrix} a_{22} & a_{23} \\ a_{32} & a_{33} \end{vmatrix},$$

$$I_3 = \begin{vmatrix} a_{11} & a_{12} & a_{13} \\ a_{21} & a_{22} & a_{23} \\ a_{31} & a_{32} & a_{33} \end{vmatrix}.$$

The polynomial in the left-hand side of the characteristic equation (of degree three for the space L_3) is called the *characteristic polynomial* of the matrix A. Since the eigenvalues of the transformation **A** are independent of the choice of basis, the roots of the characteristic equation must also be independent of the choice of the basis. As we now show, the same is true of the characteristic polynomial itself:

THEOREM. *The characteristic polynomial of a matrix A is independent of the choice of basis.*

Proof. The characteristic polynomial is the determinant of the matrix $A - \lambda E$. Under a change of basis, the matrix A goes into the matrix

$$A' = \Gamma A \Gamma^{-1}$$

(see Sec. 16.4), where Γ is the orthogonal matrix of the transformation from the old basis to the new basis. But obviously $\Gamma(\lambda E)\Gamma^{-1} = \lambda E$, and hence

$$A' - \lambda E = \Gamma A \Gamma^{-1} - \Gamma(\lambda E)\Gamma^{-1} = \Gamma(A - \lambda E)\Gamma^{-1}.$$

Therefore, by the theorem on multiplication of determinants (see Sec. 16.3),

$$|A' - \lambda E| = |\Gamma|\,|A - \lambda E|\,|\Gamma^{-1}|.$$

But $|\Gamma|\,|\Gamma^{-1}| = 1$, since the product of the determinants of a matrix and of its inverse must equal unity. It follows that

$$|A - \lambda E| = |A' - \lambda E|. \quad \blacksquare$$

Remark 1. Hence the characteristic polynomial of the matrix A can now be called the characteristic polynomial of the *transformation* **A**.

Remark 2. It follows from the invariance of the characteristic polynomial that its coefficients I_1, I_2 and I_3 are also invariant, i.e., that

$$a_{11} + a_{22} + a_{33} = a_{1'1'} + a_{2'2'} + a_{3'3'},$$

$$\begin{vmatrix} a_{11} & a_{12} \\ a_{21} & a_{22} \end{vmatrix} + \begin{vmatrix} a_{11} & a_{13} \\ a_{31} & a_{33} \end{vmatrix} + \begin{vmatrix} a_{22} & a_{23} \\ a_{32} & a_{33} \end{vmatrix}$$
$$= \begin{vmatrix} a_{1'1'} & a_{1'2'} \\ a_{2'1'} & a_{2'2'} \end{vmatrix} + \begin{vmatrix} a_{1'1'} & a_{1'3'} \\ a_{3'1'} & a_{3'3'} \end{vmatrix} + \begin{vmatrix} a_{2'2'} & a_{2'3'} \\ a_{3'2'} & a_{3'3'} \end{vmatrix},$$

$$\begin{vmatrix} a_{11} & a_{12} & a_{13} \\ a_{21} & a_{22} & a_{23} \\ a_{31} & a_{32} & a_{33} \end{vmatrix} = \begin{vmatrix} a_{1'1'} & a_{1'2'} & a_{1'3'} \\ a_{2'1'} & a_{2'2'} & a_{2'3'} \\ a_{3'1'} & a_{3'2'} & a_{3'3'} \end{vmatrix}.$$

Thus the matrix A of a linear transformation **A** of the space L_3 has three invariants.† Note that the invariance of I_1, the trace of A, and of I_3, the determinant of A, have already been proved in Secs. 11.4 and 16.4.

19.4. Turning now to the two-dimensional case, let $\mathbf{u} = \mathbf{A}\mathbf{x}$ be a linear transformation, of the plane L_2, with matrix

$$A = \begin{pmatrix} a_{11} & a_{12} \\ a_{21} & a_{22} \end{pmatrix}$$

in some orthonormal basis $\mathbf{e}_1$, $\mathbf{e}_2$. Just as in Sec. 19.2, it can be shown that the eigenvalues of the transformation are determined from the characteristic equation

$$\begin{vmatrix} a_{11} - \lambda & a_{12} \\ a_{21} & a_{22} - \lambda \end{vmatrix} = 0.$$

---

† By the same token, I_1, I_2 and I_3 are called the *invariants* of the transformation **A** itself.

The eigenvectors are then determined from the system

$$\begin{aligned}(a_{11}-\lambda)x_1+a_{12}x_2&=0,\\ a_{21}x_1+(a_{22}-\lambda)x_2&=0,\end{aligned}\tag{3'}$$

after replacing λ by the solutions λ_1 and λ_2 of the characteristic equation. As before, the polynomial

$$P(\lambda)=\lambda^2-(a_{11}+a_{22})\lambda+(a_{11}a_{22}-a_{12}a_{21})$$

appearing in the left-hand side of the characteristic equation is called the characteristic polynomial of the transformation **A**, and its coefficients

$$I_1=a_{11}+a_{22},$$

$$I_2=\begin{vmatrix}a_{11} & a_{12}\\ a_{21} & a_{22}\end{vmatrix}$$

do not depend on the choice of basis.

Next we consider the same transformations as in Examples 3 and 4 of Sec. 19.1, giving algebraic proofs of results already found geometrically:

Example 1. The transformation corresponding to rotation of the plane L_2 through the angle α has matrix

$$\begin{pmatrix}\cos\alpha & -\sin\alpha\\ \sin\alpha & \cos\alpha\end{pmatrix}$$

(cf. Example 6, p. 72). The characteristic equation of this transformation is just

$$\begin{vmatrix}\cos\alpha-\lambda & -\sin\alpha\\ \sin\alpha & \cos\alpha-\lambda\end{vmatrix}=0,$$

which implies

$$\lambda^2-2\lambda\cos\alpha+1=0.\tag{6}$$

The roots of the quadratic equation (6), equal to

$$\lambda=\cos\alpha\pm\sqrt{\cos^2\alpha-1},$$

are imaginary for all values of α between 0° and 180°. Therefore a rotation through any angle other than 0° or 180° has no real eigenvalues, and hence no real eigenvectors. But it is easy to see that the transformation in question has the conjugate complex eigenvalues

$$\lambda=\cos\alpha\pm i\sin\alpha,$$

with corresponding complex eigenvectors that can be determined from the system

$$\begin{aligned}\mp ix_1\sin\alpha-x_2\sin\alpha&=0,\\ x_1\sin\alpha\mp ix_2\sin\alpha&=0.\end{aligned}$$

Since $\alpha\neq 0°, 180°$, this implies

$$x_2=\mp ix_1.$$

Setting $x_1 = 1$, we get the conjugate complex eigenvectors $\mathbf{a}_1 = (1, -i)$ and $\mathbf{a}_2 = (1, i)$. Note that $\mathbf{a}_1 \cdot \mathbf{a}_1 = \mathbf{a}_2 \cdot \mathbf{a}_2 = 0$, so that $\mathbf{a}_1$ and $\mathbf{a}_2$ both have "zero length."†

Example 2. The shift considered in Example 4, p. 108 has matrix

$$\begin{pmatrix} 1 & k \\ 0 & 1 \end{pmatrix}$$

and characteristic equation

$$\begin{vmatrix} 1-\lambda & k \\ 0 & 1-\lambda \end{vmatrix} = 0,$$

i.e.,

$$(1-\lambda)^2 = 0,$$

with roots

$$\lambda_1 = \lambda_2 = 1.$$

The corresponding eigenvectors are determined from the system

$$(1-1)x_1 + kx_2 = 0,$$
$$0 \cdot x_1 + (1-1)x_2 = 0,$$

which implies $x_2 = 0$, i.e., every eigenvector lies on the x_1-axis and has eigenvalue 1, as already noted.

We conclude this section with two numerical examples.

Example 3. Find the eigenvectors and eigenvalues of the linear transformation $\mathbf{u} = \mathbf{A}\mathbf{x}$ which has the component form

$$u_1 = 3x_1 + 4x_2,$$
$$u_2 = 5x_1 + 2x_2,$$

relative to an orthonormal basis $\mathbf{e}_1, \mathbf{e}_2$.

Solution. Solving the characteristic equation

$$\begin{vmatrix} 3-\lambda & 4 \\ 5 & 2-\lambda \end{vmatrix} = 0,$$

or

$$\lambda^2 - 5\lambda - 14 = 0,$$

we find the eigenvalues

$$\lambda_1 = -2, \qquad \lambda_2 = 7.$$

---

† Actually, in the case where L_2 is complex, the appropriate defintion of the scalar product of two vectors $\mathbf{x} = x_i\mathbf{e}_i$, $\mathbf{y} = y_i\mathbf{e}_i$ is

$$\mathbf{x} \cdot \mathbf{y} = x_1\bar{y}_1 + x_2\bar{y}_2$$

(the overbar denotes the complex conjugate), rather than

$$\mathbf{x} \cdot \mathbf{y} = x_1y_1 + x_2y_2.$$

The lengths of $\mathbf{a}_1$ and $\mathbf{a}_2$ are then both equal to $\sqrt{2}$ rather than 0.

For $\lambda_1 = -2$ the eigenvector can be determined from the system

$$5x_1 + 4x_2 = 0,$$
$$5x_1 + 4x_2 = 0,$$

which implies

$$\frac{x_1}{x_2} = -\frac{4}{5}.$$

Similarly, for $\lambda_2 = 9$ we have

$$-4x_1 + 4x_2 = 0,$$
$$5x_1 - 5x_2 = 0,$$

which implies

$$\frac{x_1}{x_2} = 1.$$

Thus the eigenvectors of the transformation **A** are

$$\mathbf{a}_1 = (4, -5), \qquad \mathbf{a}_2 = (1, 1),$$

and all vectors collinear with $\mathbf{a}_1$ and $\mathbf{a}_2$.

Example 4. Find the real eigenvectors and eigenvalues of the linear transformation $\mathbf{u} = \mathbf{A}\mathbf{x}$, which has the component form

$$\begin{aligned} u_1 &= 4x_1 - 5x_2 + 7x_3, \\ u_2 &= x_1 - 4x_2 + 9x_3, \\ u_3 &= -4x_1 \qquad\quad + 5x_3 \end{aligned}$$

relative to an orthonormal basis $\mathbf{e}_1, \mathbf{e}_2, \mathbf{e}_3$.

Solution. The characteristic equation of **A** is

$$\begin{vmatrix} 4-\lambda & -5 & 7 \\ 1 & -4-\lambda & 9 \\ -4 & 0 & 5-\lambda \end{vmatrix} = 0,$$

or

$$\lambda^3 - 5\lambda^2 + 17\lambda - 13 = 0,$$

with roots

$$\lambda_1 = 1, \qquad \lambda_2 = 2 + 3i, \qquad \lambda_3 = 2 - 3i.$$

The eigenvector corresponding to the unique real eigenvalue $\lambda_1 = 1$ can be found from the system

$$\begin{aligned} 3x_1 - 5x_2 + 7x_3 &= 0, \\ x_1 - 5x_2 + 9x_3 &= 0, \\ -4x_1 \qquad\quad + 4x_3 &= 0. \end{aligned}$$

It follows from the last equation of this system that $x_1 = x_3$, and then from the first two that

$$\frac{x_1}{x_2} = \frac{1}{2}.$$

Thus the only real eigenvector of the transformation **A** is **a** = (1, 2, 1), or any vector collinear with **a**.

PROBLEMS

1. Prove that every vector of the space L_1 is an eigenvector of every linear transformation of L_1.

2. Find the eigenvectors and eigenvalues of the following linear transformations of the space L_3:

a) $\mathbf{u} = (\mathbf{a}\cdot\mathbf{x})\mathbf{b}$;

b) $\mathbf{u} = \mathbf{a} \times \mathbf{x}$;

c) $\mathbf{u} = (\mathbf{x}\cdot\boldsymbol{\omega})\boldsymbol{\omega} + [\mathbf{x} - (\mathbf{x}\cdot\boldsymbol{\omega})\boldsymbol{\omega}] \cos\alpha + \boldsymbol{\omega} \times \mathbf{x} \sin\alpha$, where $\boldsymbol{\omega}$ is a unit vector;

d) $\mathbf{u} = (\mathbf{a}\cdot\mathbf{x})\mathbf{a} + (\mathbf{b}\cdot\mathbf{x})\mathbf{b}$, where $|\mathbf{a}| = |\mathbf{b}|$;

e) $\mathbf{u} = (\mathbf{a}\cdot\mathbf{x})\mathbf{a} + (\mathbf{b}\cdot\mathbf{x})\mathbf{b} + (\mathbf{c}\cdot\mathbf{x})\mathbf{c}$, where $|\mathbf{a}| = |\mathbf{b}| = |\mathbf{c}|$ and $\mathbf{a}\cdot\mathbf{b} = \mathbf{b}\cdot\mathbf{c} = \mathbf{a}\cdot\mathbf{c}$.

3. Find the eigenvectors and eigenvalues of the linear transformation

a) Carrying the vectors $\mathbf{e}_1, \mathbf{e}_2, \mathbf{e}_3$ into the vectors $\mathbf{e}_2, \mathbf{e}_3, \mathbf{e}_1$;

b) Carrying the vectors $\mathbf{e}_1, \mathbf{e}_2, \mathbf{e}_3$ into the vectors $\mathbf{e}_2 + \mathbf{e}_3, \mathbf{e}_3 + \mathbf{e}_1, \mathbf{e}_1 + \mathbf{e}_2$.

4. Find the eigenvectors and eigenvalues of the linear transformations of the plane L_2 and the space L_3 with the following matrices (in some orthonormal basis):

a) $\begin{pmatrix} 2 & 1 \\ 2 & 3 \end{pmatrix}$; b) $\begin{pmatrix} 2 & -1 & -1 \\ 0 & -1 & 0 \\ 0 & 2 & 1 \end{pmatrix}$; c) $\begin{pmatrix} 1 & 2 & 3 \\ 2 & 1 & 3 \\ 3 & 3 & 6 \end{pmatrix}$;

d) $\begin{pmatrix} a & -a^2 & a^3 \\ 1 & 0 & 0 \\ 0 & 1 & 0 \end{pmatrix}$; e) $\begin{pmatrix} a & 0 & 0 \\ 1 & a & 0 \\ 0 & 1 & a \end{pmatrix}$; f) $\begin{pmatrix} a_1 & b_1 & c_1 \\ 0 & b_2 & c_2 \\ 0 & 0 & c_3 \end{pmatrix}$.

5. Prove that

a) The characteristic equations of the linear transformation **A** and of the transformation **A*** adjoint to **A** are identical;

b) If **x** is an eigenvector of the transformation **A** with eigenvalue λ_1 and of the transformation **A*** with eigenvalue λ_2, then $\lambda_1 = \lambda_2$.

6. Let λ_1, λ_2 and λ_3 be eigenvalues of a linear transformation **A**. Prove that

$$\lambda_1 + \lambda_2 + \lambda_3 = I_1, \qquad \lambda_1\lambda_2 + \lambda_2\lambda_3 + \lambda_3\lambda_1 = I_2, \qquad \lambda_1\lambda_2\lambda_3 = I_3.$$

7. Using the result of the preceding problem, prove that the eigenvalues of a transformation **A** are all nonzero if and only if the transformation **A** is nonsingular.

8. Prove that the eigenvalues of the inverse transformation $\mathbf{A}^{-1}$ are the reciprocals of the eigenvalues of the original transformation **A**.

9. Prove that the transformations **AB** and **BA** both have the same characteristic polynomial.

10. Prove that the proper rotation **A** ($\neq$ **E**) of the space L_3 with matrix

$$A = \begin{pmatrix} a_{11} & a_{12} & a_{13} \\ a_{21} & a_{22} & a_{23} \\ a_{31} & a_{32} & a_{33} \end{pmatrix}, \qquad |A| = 1$$

in some orthonormal basis is equivalent to a rotation α about some fixed axis l. Find α and the direction of l.

11. Find the angle α and the direction of l figuring in the preceding problem if

$$A = \begin{pmatrix} \frac{11}{15} & \frac{2}{15} & \frac{2}{3} \\ \frac{2}{15} & \frac{14}{15} & -\frac{1}{3} \\ -\frac{2}{3} & \frac{1}{3} & \frac{2}{3} \end{pmatrix}.$$

12. The transformation **A** with matrix

$$\begin{pmatrix} \frac{1}{2} & -\frac{1}{\sqrt{2}} & \frac{1}{2} \\ \frac{1}{2} & \frac{1}{\sqrt{2}} & -\frac{1}{2} \\ \frac{1}{\sqrt{2}} & 0 & \frac{1}{\sqrt{2}} \end{pmatrix}$$

corresponds to a proper rotation through an angle α about some axis. Find the transformation **B** which corresponds to a rotation through the angle $-\alpha$ about the same axis.

13. Prove that if **x** is an eigenvector of the transformation **A** corresponding to the eigenvalue λ, then **x** is also an eigenvector of the transformation $\mathbf{A}^2$ corresponding to the eigenvalue λ^2.

14. Prove that if the transformation $\mathbf{A}^2$ has an eigenvector with a nonnegative eigenvalue μ^2, then the transformation **A** also has an eigenvector.

15. Prove that if the characteristic equation of a linear transformation **A** of the space L_3 has two conjugate complex roots, then there is a plane (called an *invariant plane*) which is carried into itself by the transformation **A**. Find this plane for the transformation considered in Example 4, p. 114.

16. Let $C[a, b]$ be the space of all functions continuous in the interval $[a, b]$, and let **A** be the transformation which consists in multiplying a function $f(t) \in C[a, b]$ by the independent variable t. Prove that **A** has no eigenvalues.

17. Prove that the differentiation operator in the space $C[a, b]$ has infinitely many eigenvalues.

18. Find the eigenvectors and eigenvalues of the differentiation operator in the space of all polynomials of degree not exceeding n.

20. The Case of Distinct Eigenvalues

In the case of distinct eigenvalues we have the following elegant results:

THEOREM 1. *Let* **A** *be a linear transformation of the space* L_3 *whose characteristic equation has three distinct real roots* $\lambda_1, \lambda_2, \lambda_3$, *and let* $\mathbf{a}_1, \mathbf{a}_2, \mathbf{a}_3$ *be the eigenvectors with eigenvalues* $\lambda_1, \lambda_2, \lambda_3$, *respectively. Then the vectors* $\mathbf{a}_1, \mathbf{a}_2, \mathbf{a}_3$ *are linearly independent.*

Proof. By hypothesis,

$$\mathbf{Aa}_1 = \lambda_1\mathbf{a}_1, \qquad \mathbf{Aa}_2 = \lambda_2\mathbf{a}_2, \qquad \mathbf{Aa}_3 = \lambda_3\mathbf{a}_3.$$

Suppose two of the vectors $\mathbf{a}_1, \mathbf{a}_2, \mathbf{a}_3$, say $\mathbf{a}_1$ and $\mathbf{a}_2$, are connected by a linear relation

$$\alpha_1\mathbf{a}_1 + \alpha_2\mathbf{a}_2 = \mathbf{0} \tag{1}$$

($\mathbf{a}_1$ and $\mathbf{a}_2$ are both nonzero, being eigenvectors). Applying the transformation **A** to (1), we get

$$\alpha_1\mathbf{Aa}_1 + \alpha_2\mathbf{Aa}_2 = \mathbf{0}$$

or

$$\alpha_1\lambda_1\mathbf{a}_1 + \alpha_2\lambda_2\mathbf{a}_2 = \mathbf{0}. \tag{2}$$

Multiplying (1) first by $-\lambda_1$ and then by $-\lambda_2$, and adding each of the resulting equations to (2), we find that

$$\alpha_2(\lambda_2 - \lambda_1)\mathbf{a}_2 = \mathbf{0}, \qquad \alpha_1(\lambda_1 - \lambda_2)\mathbf{a}_1 = \mathbf{0},$$

which implies

$$\alpha_1 = \alpha_2 = 0,$$

since $\lambda_1 \neq \lambda_2$. It follows that $\mathbf{a}_1$ and $\mathbf{a}_2$ are linearly independent.

We have just proved the linear independence of any two of the vectors $\mathbf{a}_1, \mathbf{a}_2, \mathbf{a}_3$. To prove that all three vectors $\mathbf{a}_1, \mathbf{a}_2, \mathbf{a}_3$ are linearly independent, suppose to the contrary that $\mathbf{a}_1, \mathbf{a}_2, \mathbf{a}_3$ are linearly dependent, so that

$$\alpha_1\mathbf{a}_1 + \alpha_2\mathbf{a}_2 + \alpha_3\mathbf{a}_3 = \mathbf{0}, \tag{3}$$

where $\alpha_1 \neq 0$, say. Applying the transformation **A** to (3), we get

$$\alpha_1\mathbf{Aa}_1 + \alpha_2\mathbf{Aa}_2 + \alpha_3\mathbf{Aa}_3 = \mathbf{0}$$

or

$$\alpha_1\lambda_1\mathbf{a}_1 + \alpha_2\lambda_2\mathbf{a}_2 + \alpha_3\lambda_3\mathbf{a}_3 = \mathbf{0}. \tag{4}$$

Multiplying (3) by $-\lambda_3$ and adding the resulting equation to (4), we then get

$$\alpha_1(\lambda_1 - \lambda_3)\mathbf{a}_1 + \alpha_2(\lambda_2 - \lambda_3)\mathbf{a}_2 = \mathbf{0},$$

from which it follows that $\mathbf{a}_1$ and $\mathbf{a}_2$ are linearly dependent, since $\alpha_1 \neq 0$,

$\lambda_1 \neq \lambda_3$. This contradiction shows that $\mathbf{a}_1, \mathbf{a}_2, \mathbf{a}_3$ are in fact linearly independent. ∎

THEOREM 2. *Let* **A** *be the same as in Theorem* 1, *with eigenvalues* $\lambda_1, \lambda_2, \lambda_3$ *and corresponding eigenvectors* $\mathbf{a}_1, \mathbf{a}_2, \mathbf{a}_3$. *Then the matrix of* **A** *takes the particularly simple "diagonal form"*

$$A = \begin{pmatrix} \lambda_1 & 0 & 0 \\ 0 & \lambda_2 & 0 \\ 0 & 0 & \lambda_3 \end{pmatrix} \tag{5}$$

in the basis consisting of the eigenvectors $\mathbf{a}_1, \mathbf{a}_2, \mathbf{a}_3$.

Proof. Being linearly independent, the vectors $\mathbf{a}_1, \mathbf{a}_2, \mathbf{a}_3$ can be chosen as a basis. (Note that this basis is in general not orthonormal or even orthogonal, but the considerations of Sec. 14 apply equally well to the nonorthogonal case.) Given any vector $\mathbf{x}$, suppose

$$\mathbf{x} = x_1\mathbf{a}_1 + x_2\mathbf{a}_2 + x_3\mathbf{a}_3,$$

and let

$$\mathbf{u} = \mathbf{A}\mathbf{x} = u_1\mathbf{a}_1 + u_2\mathbf{a}_2 + u_3\mathbf{a}_3.$$

Then

$$\begin{aligned} \mathbf{u} = \mathbf{A}(x_1\mathbf{a}_1 + x_2\mathbf{a}_2 + x_3\mathbf{a}_3) &= x_1\mathbf{A}\mathbf{a}_1 + x_2\mathbf{A}\mathbf{a}_2 + x_3\mathbf{A}\mathbf{a}_3 \\ &= x_1\lambda_1\mathbf{a}_1 + x_2\lambda_2\mathbf{a}_2 + x_3\lambda_3\mathbf{a}_3, \end{aligned}$$

so that

$$u_1 = \lambda_1 x_1, \qquad u_2 = \lambda_2 x_2, \qquad u_3 = \lambda_3 x_3.$$

But then **A** has the matrix (5) in the basis $\mathbf{a}_1, \mathbf{a}_2, \mathbf{a}_3$. ∎

It is clear from Theorem 2 that the eigenvectors play an important role in the theory of linear transformations. In fact, if there exists a basis consisting of eigenvectors, then the transformation **A** has its simplest "component representation" in this basis, with a matrix involving only the eigenvalues of **A**.

Remark 1. The converse of Theorem 2 states that *if a linear transformation* **A** *has the diagonal matrix* (5) *in some basis* $\mathbf{a}_1, \mathbf{a}_2, \mathbf{a}_3$, *where* $\lambda_1, \lambda_2, \lambda_3$ *are distinct real numbers, then the vectors* $\mathbf{a}_1, \mathbf{a}_2, \mathbf{a}_3$ *are eigenvectors of* **A**. This proposition has in effect already been proved in Example 5, p. 108.

Remark 2. Theorem 2 has an obvious analogue for the case of the plane L_2, i.e., *if* **A** *is a linear transformation of* L_2 *whose characteristic equation has two distinct real roots* λ_1 *and* λ_2, *then the matrix of* **A** *takes the diagonal form*

$$A = \begin{pmatrix} \lambda_1 & 0 \\ 0 & \lambda_2 \end{pmatrix}$$

in the basis consisting of the corresponding eigenvectors $\mathbf{a}_1$ *and* $\mathbf{a}_2$ ($\mathbf{a}_1$ *and* $\mathbf{a}_2$ are noncollinear but not necessarily orthogonal, by the analogue of Theorem 1).

Example 1. A linear transformation **A** of the plane L_2 has the matrix

$$A = \begin{pmatrix} 3 & 4 \\ 5 & 2 \end{pmatrix}$$

in some orthonormal basis. Reduce this matrix to diagonal form by making a suitable transformation to a new basis.

Solution. As shown in Example 3, p. 113, the transformation **A** has eigenvalues $\lambda_1 = -2$, $\lambda_2 = 7$, with corresponding eigenvectors $\mathbf{a}_1 = (4, -5)$, $\mathbf{a}_2 = (1, 1)$. In the (nonorthogonal) basis consisting of the vectors $\mathbf{a}_1$ and $\mathbf{a}_2$, the matrix of **A** takes the diagonal form

$$A' = \begin{pmatrix} -2 & 0 \\ 0 & 9 \end{pmatrix}.$$

Example 2. A linear transformation **A** of the space L_3 has the matrix

$$A = \begin{pmatrix} 5 & -3 & 2 \\ 6 & -4 & 4 \\ 4 & -4 & 5 \end{pmatrix}$$

in some orthonormal basis. Reduce this matrix to diagonal form by making a suitable transformation to a new basis.

Solution. The characteristic equation of **A** is

$$\begin{vmatrix} 5-\lambda & -3 & 2 \\ 6 & -4-\lambda & 4 \\ 4 & -4 & 5-\lambda \end{vmatrix} = 0$$

or

$$\lambda^3 - 6\lambda^2 + 11\lambda - 6 = 0,$$

with distinct real roots $\lambda_1 = 1, \lambda_2 = 2, \lambda_3 = 3$. We now determine the corresponding eigenvectors.

1) For $\lambda_1 = 1$ we have

$$\begin{aligned} 4x_1 - 3x_2 + 2x_3 &= 0, \\ 6x_1 - 5x_2 + 4x_3 &= 0, \\ 4x_1 - 4x_2 + 4x_3 &= 0. \end{aligned}$$

It follows from the third equation that $x_2 = x_1 + x_3$, and the first two equations then imply $x_1 = x_3$, so that $\mathbf{a}_1 = (1, 2, 1)$.

2) For $\lambda_2 = 2$ we have

$$\begin{aligned} 3x_1 - 3x_2 + 2x_3 &= 0, \\ 6x_1 - 6x_2 + 4x_3 &= 0, \\ 4x_1 - 4x_2 + 3x_3 &= 0. \end{aligned}$$

It follows from the first and third equations that $x_1 = x_2$, and the second equation then implies $x_3 = 0$, so that $\mathbf{a}_2 = (1, 1, 0)$.

3) For $\lambda_3 = 3$ we have

$$2x_1 - 3x_2 + 2x_3 = 0,$$
$$6x_1 - 7x_2 + 4x_3 = 0,$$
$$4x_1 - 4x_2 + 2x_3 = 0.$$

It follows from the last equation that $x_3 = -2x_1 + 2x_2$, and the first two equations then imply $x_2 = 2x_1$, so that $\mathbf{a}_3 = (1, 2, 2)$.

Going over now to the basis $\mathbf{a}_1, \mathbf{a}_2, \mathbf{a}_3$, we "diagonalize" the transformation $\mathbf{A}$, i.e., we reduce the matrix of $\mathbf{A}$ to the diagonal form

$$A' = \begin{pmatrix} 1 & 0 & 0 \\ 0 & 2 & 0 \\ 0 & 0 & 3 \end{pmatrix}.$$

The transformation $\mathbf{A}$ clearly carries the vector $\mathbf{x} = x_1\mathbf{a}_1 + x_2\mathbf{a}_2 + x_3\mathbf{a}_3$ into the vector $\mathbf{u} = \mathbf{A}\mathbf{x} = x_1\mathbf{a}_1 + 2x_2\mathbf{a}_2 + 3x_3\mathbf{a}_3$.

PROBLEMS

1. Prove the result stated in Remark 2.

2. Prove that the matrix of each of the linear transformations considered in Examples 1, 2, 5, p. 108 and Probs. 2d, 4a, 4b, 4c, 4f, p. 115 can be reduced to diagonal form by going over to a new basis. In each case find the appropriate diagonal basis and the corresponding basis.

3. The matrix of a linear transformation $\mathbf{A}$ is of the form

$$\begin{pmatrix} 0 & 0 & \alpha_1 \\ 0 & \alpha_2 & 0 \\ \alpha_3 & 0 & 0 \end{pmatrix}$$

in some orthonormal basis. When does $\mathbf{A}$ have three distinct real eigenvalues? Find the eigenvectors in this case.

4. Prove that if $\mathbf{x}$ and $\mathbf{y}$ are eigenvectors of a linear transformation $\mathbf{A}$ with *distinct* eigenvalues λ_1 and λ_2, then the vector $\alpha\mathbf{x} + \beta\mathbf{y}$ ($\alpha \neq 0, \beta \neq 0$) cannot be an eigenvector of $\mathbf{A}$.

5. Using the result of the preceding problem, show that if every vector of the space L_3 is an eigenvector of a linear transformation $\mathbf{A}$, then $\mathbf{A} = \lambda\mathbf{E}$, i.e., $\mathbf{A}$ is a homothetic transformation of L_3.

6. Prove that the matrix of a proper orthogonal transformation $\mathbf{A}$ of the plane L_2 (see the relevant footnote on p. 103) can be reduced to the form

$$\begin{pmatrix} \cos\alpha & -\sin\alpha \\ \sin\alpha & \cos\alpha \end{pmatrix}$$

in some orthonormal basis, while the matrix of an improper orthogonal transformation can be reduced to the form

$$\begin{pmatrix} 1 & 0 \\ 0 & -1 \end{pmatrix}.$$

Interpret the first transformation as a rotation and the second as a reflection.

7. Prove that the matrix of a proper orthogonal transformation **A** of the space L_3 (see Remark 2, p. 103) can be reduced to the form

$$\begin{pmatrix} 1 & 0 & 0 \\ 0 & \cos\alpha & -\sin\alpha \\ 0 & \sin\alpha & \cos\alpha \end{pmatrix}$$

in some orthonormal basis, while the matrix of an improper orthogonal transformation can be reduced to the form

$$\begin{pmatrix} -1 & 0 & 0 \\ 0 & \cos\alpha & -\sin\alpha \\ 0 & \sin\alpha & \cos\alpha \end{pmatrix}.$$

Interpret the first transformation as a rotation and the second as the product of a rotation and a reflection.

21. Matrix Polynomials and the Hamilton–Cayley Theorem

21.1. In Chapter 3 we showed how linear transformations of the plane L_2 and of the space L_3 (and the corresponding square matrices of orders two and three) are added and multiplied both by numbers and by one another. In this regard, it should be noted that if **A** is a linear transformation, then

$$\mathbf{A}^n = \underbrace{\mathbf{A}\cdot\mathbf{A}\cdots\mathbf{A}}_{n \text{ times}}$$

(by definition) for a positive integer, while

$$\mathbf{A}^0 = \mathbf{E},$$

where **E** is the identity transformation. Moreover,

$$\mathbf{A}^n = (\mathbf{A}^{-1})^{-n}$$

if **A** is nonsingular and n is a negative integer.

Now let

$$P(\lambda) = a_0\lambda^m + a_1\lambda^{m-1} + \cdots + a_{m-1}\lambda + a_m$$

be a polynomial in the variable λ. Then the expression

$$P(\mathbf{A}) = a_0\mathbf{A}^m + a_1\mathbf{A}^{m-1} + \cdots + a_{m-1}\mathbf{A} + a_m\mathbf{E}$$

is called a *polynomial in the transformation* **A**. Clearly $P(\mathbf{A})$ is a linear transformation, like **A** itself. If A is the matrix of the transformation **A** in some basis, then the matrix of $P(\mathbf{A})$ is the "matrix polynomial"

$$P(A) = a_0A^m + a_1A^{m-1} + \cdots + a_{m-1}A + a_mE,$$

where E is the unit matrix. In fact, the transformation $P(\mathbf{A})$ is obtained from $\mathbf{A}$ by the operations of multiplication and addition. But these operations on linear transformations lead to the same operations on the corresponding matrices.

All the algebraic rules valid for polynomials in one variable continue to hold for polynomials in linear transformations and for the corresponding matrix polynomials. For example,

$$(\mathbf{A} + \mathbf{E})^2 = \mathbf{A}^2 + 2\mathbf{A} + \mathbf{E},$$

$$(\mathbf{A} + \mathbf{E})^3 = \mathbf{A}^3 + 3\mathbf{A}^2 + 3\mathbf{A} + \mathbf{E},$$

$$\mathbf{A}^2 - \mathbf{E} = (\mathbf{A} + \mathbf{E})(\mathbf{A} - \mathbf{E}),$$

and so on. Note that two polynomials $P(\mathbf{A})$ and $Q(\mathbf{A})$ in the same linear transformation $\mathbf{A}$ always commute:

$$P(\mathbf{A})Q(\mathbf{A}) = Q(\mathbf{A})P(\mathbf{A}).$$

21.2. A linear transformation $\mathbf{A}$ is called a *root* of the polynomial $P(\lambda)$ if the substitution $\lambda = \mathbf{A}$ reduces $P(\lambda)$ to the null transformation, i.e., if $P(\mathbf{A}) = \mathbf{N}$.

THEOREM (***Hamilton–Cayley***). *Let* $\mathbf{A}$ *be a linear transformation with characteristic polynomial* $P(\lambda)$ *and distinct real eigenvalues* $\lambda_1, \lambda_2, \lambda_3$. *Then* $\mathbf{A}$ *is a root of its own characteristic polynomial, i.e.,*

$$P(\mathbf{A}) = \mathbf{N}. \tag{1}$$

Proof. Since

$$P(\lambda) = (\lambda - \lambda_1)(\lambda - \lambda_2)(\lambda - \lambda_3),$$

we have

$$P(\mathbf{A}) = (\mathbf{A} - \lambda_1\mathbf{E})(\mathbf{A} - \lambda_2\mathbf{E})(\mathbf{A} - \lambda_3\mathbf{E}),$$

where the product on the right does not depend on the order of the factors. To prove (1), we must show that $P(\mathbf{A})$ carries every vector $\mathbf{x}$ into the zero vector, i.e., that $P(\mathbf{A})\mathbf{x} = \mathbf{0}$ for every $\mathbf{x}$. Let $\mathbf{a}_1, \mathbf{a}_2, \mathbf{a}_3$ be the eigenvectors of $\mathbf{A}$ corresponding to the eigenvalues $\lambda_1, \lambda_2, \lambda_3$, so that

$$\mathbf{A}\mathbf{a}_1 = \lambda_1\mathbf{a}_1, \qquad \mathbf{A}\mathbf{a}_2 = \lambda_2\mathbf{a}_2, \qquad \mathbf{A}\mathbf{a}_3 = \lambda_3\mathbf{a}_3.$$

By Theorem 1, p. 117, the vectors $\mathbf{a}_1, \mathbf{a}_2, \mathbf{a}_3$ are linearly independent and hence form a basis, so that any vector $\mathbf{x} \in L_3$ can be written as a linear combination

$$\mathbf{x} = x_1\mathbf{a}_1 + x_2\mathbf{a}_2 + x_3\mathbf{a}_3.$$

We then have

$$P(\mathbf{A})\mathbf{x} = x_1P(\mathbf{A})\mathbf{a}_1 + x_2P(\mathbf{A})\mathbf{a}_2 + x_3P(\mathbf{A})\mathbf{a}_3.$$

But

$$\begin{aligned} P(\mathbf{A})\mathbf{a}_1 &= (\mathbf{A} - \lambda_2\mathbf{E})(\mathbf{A} - \lambda_3\mathbf{E})(\mathbf{A} - \lambda_1\mathbf{E})\mathbf{a}_1 \\ &= (\mathbf{A} - \lambda_2\mathbf{E})(\mathbf{A} - \lambda_3\mathbf{E})(\mathbf{A}\mathbf{a}_1 - \lambda_1\mathbf{E}\mathbf{a}_1) \\ &= (\mathbf{A} - \lambda_2\mathbf{E})(\mathbf{A} - \lambda_3\mathbf{E})(\lambda_1\mathbf{a}_1 - \lambda_1\mathbf{a}_1) = \mathbf{0}, \end{aligned}$$

and similarly

$$P(\mathbf{A})\mathbf{a}_2 = \mathbf{0}, \qquad P(\mathbf{A})\mathbf{a}_3 = \mathbf{0}.$$

It follows that

$$P(\mathbf{A})\mathbf{x} = \mathbf{0}$$

for every **x**. ▮

Remark. It can be shown that the Hamilton–Cayley theorem remains true even if some or all of the eigenvalues $\lambda_1, \lambda_2, \lambda_3$ coincide.†

21.3. If $P(\lambda)$ is the characteristic polynomial of the linear transformation **A**, then

$$P(\lambda) = \lambda^3 - I_1\lambda^2 + I_2\lambda - I_3,$$

where I_1, I_2 and I_3 are the invariants of **A** (see Sec. 19.3). It follows from the Hamilton–Cayley theorem that the matrices E, A, A^2, A^3 are linearly dependent, since

$$A^3 - I_1A^2 + I_2A - I_3E = N, \tag{2}$$

where N is the null matrix. This also implies that any four consecutive powers $A^k, A^{k+1}, A^{k+2}, A^{k+3}$ of the matrix A are linearly dependent, as we see at once by multiplying (2) by A^k. We can now write a new expression for the inverse matrix A^{-1} of a nonsingular matrix A. In fact, multiplying (2) by A^{-1}, we get

$$A^2 - I_1A + I_2E - I_3A^{-1} = N.$$

But $I_3 = |A| \neq 0$, since A is nonsingular, and hence

$$A^{-1} = \frac{1}{I_3}(A^2 - I_1A + I_2E). \tag{3}$$

PROBLEMS

1. Find $\varphi(A)$ if

$$\varphi(\lambda) = -2 - 5\lambda + 3\lambda^2, \qquad A = \begin{pmatrix} 1 & 2 \\ 3 & 1 \end{pmatrix}.$$

2. Prove by direct substitution that the matrix

$$\begin{pmatrix} a & b \\ c & d \end{pmatrix}$$

† See e.g., G. E. Shilov, *op. cit.*, Sec. 6.52.

satisfies its own characteristic equation

$$\lambda^2 - (a + d)\lambda + ad - bc = 0.$$

3. Let $f(\mathbf{A})$ be a polynomial in a linear transformation **A**. Prove that

a) The eigenvectors of the linear transformation $f(\mathbf{A})$ coincide with those of **A** itself;

b) If λ is an eigenvalue of **A**, then $f(\lambda)$ is an eigenvalue of $f(\mathbf{A})$.

4. Given any vector $\mathbf{a} \in L_3$ and any linear transformation **A**, let

$$\mathbf{a}_1 = \mathbf{A}\mathbf{a}, \qquad \mathbf{a}_2 = \mathbf{A}^2\mathbf{a}, \qquad \mathbf{a}_3 = \mathbf{A}^3\mathbf{a}.$$

Prove that

a) $\mathbf{a}_3 = I_1\mathbf{a}_2 - I_2\mathbf{a}_1 + I_3\mathbf{a}$;

b) If the vectors $\mathbf{a}, \mathbf{a}_1, \mathbf{a}_2$ are linearly dependent and the vectors $\mathbf{a}, \mathbf{a}_1$ are noncollinear, then the plane determined by **a** and $\mathbf{a}_1$ is carried into itself by **A**, i.e., is invariant under **A**.

Assuming that the vectors $\mathbf{a}, \mathbf{a}_1, \mathbf{a}_2$ are linearly independent, choose them as basis vectors and find the matrix of **A** in this basis.

5. Prove that the relation $AB - BA = E$ cannot hold for any choice of the matrices A and B.

6. Use formula (3) and its analogue for the plane L_2 to find the inverses of the matrices figuring in Prob. 1, p. 97.

22. Eigenvectors of a Symmetric Transformation

Let **A** be a symmetric linear transformation of the space L_3.† Then

$$(\mathbf{x}, \mathbf{A}\mathbf{y}) = (\mathbf{y}, \mathbf{A}\mathbf{x})$$

for arbitrary vectors **x** and **y** (see p. 81). Moreover, a linear transformation **A** is symmetric if and only if it has a symmetric matrix in every orthonormal basis (see Sec. 15.4).

We now prove a number of theorems on the eigenvectors and eigenvalues of a symmetric linear transformation **A**. These theorems will allow us to solve the problem of finding the simplest form of the matrix of **A** and of interpreting **A** geometrically.

THEOREM 1. *Let* **A** *be a symmetric linear transformation. Then the eigenvectors of* **A** *corresponding to distinct eigenvalues are orthogonal.*

Proof. Given two distinct eigenvalues λ_1 and λ_2 of **A**, let $\mathbf{a}_1$ and $\mathbf{a}_2$ be the corresponding eigenvectors, so that

$$\mathbf{A}\mathbf{a}_1 = \lambda_1\mathbf{a}_1,$$
$$\mathbf{A}\mathbf{a}_2 = \lambda_2\mathbf{a}_2.$$

---

† Or, more generally, of the plane L_2 or the space L_n (defined in just the same way as on p. 81).

Taking the scalar product of the first equation with $\mathbf{a}_2$ and of the second equation with $\mathbf{a}_1$, we get

$$(\mathbf{a}_2, \mathbf{A}\mathbf{a}_1) = \lambda_1(\mathbf{a}_1, \mathbf{a}_2),$$
$$(\mathbf{a}_1, \mathbf{A}\mathbf{a}_2) = \lambda_2(\mathbf{a}_1, \mathbf{a}_2).$$

But $(\mathbf{a}_2, \mathbf{A}\mathbf{a}_1) = (\mathbf{a}_1, \mathbf{A}\mathbf{a}_2)$ by the symmetry of $\mathbf{A}$, and hence

$$\lambda_1(\mathbf{a}_1, \mathbf{a}_2) = \lambda_2(\mathbf{a}_1, \mathbf{a}_2)$$

or

$$(\lambda_1 - \lambda_2)(\mathbf{a}_1, \mathbf{a}_2) = 0.$$

Since $\lambda_1 \neq \lambda_2$, it follows that

$$(\mathbf{a}_1, \mathbf{a}_2) = 0,$$

i.e., $\mathbf{a}_1$ and $\mathbf{a}_2$ are orthogonal. ▮

THEOREM 2. *Let* $\mathbf{a}$ *be an eigenvector of a symmetric linear transformation* $\mathbf{A}$ *and let* $\mathbf{x}$ *be a vector orthogonal to* $\mathbf{a}$. *Then the vector* $\mathbf{A}\mathbf{x}$ *is also orthogonal to* $\mathbf{a}$.

Proof. We have

$$(\mathbf{a}, \mathbf{x}) = 0$$

since $\mathbf{x}$ is orthogonal to $\mathbf{a}$, and

$$\mathbf{A}\mathbf{a} = \lambda\mathbf{a}$$

for some λ since $\mathbf{a}$ is an eigenvector of $\mathbf{A}$. Therefore

$$(\mathbf{a}, \mathbf{A}\mathbf{x}) = (\mathbf{x}, \mathbf{A}\mathbf{a}) = (\mathbf{x}, \lambda\mathbf{a}) = \lambda(\mathbf{a}, \mathbf{x}) = 0,$$

i.e., $\mathbf{A}\mathbf{x}$ is orthogonal to $\mathbf{a}$. ▮

Remark 1. The dimension of the underlying linear space plays no role in the proofs of Theorems 1 and 2. Therefore both theorems are valid for any space L_n, in particular for the plane L_2.

Remark 2. For the plane L_2 it follows from Theorem 2 that if $\mathbf{a}$ is an eigenvector of $\mathbf{A}$, then every vector orthogonal to $\mathbf{a}$ is also an eigenvector of $\mathbf{A}$. For the space L_3 it follows from Theorem 2 that if $\mathbf{a}$ is an eigenvector of $\mathbf{A}$ and if Π is the plane perpendicular to $\mathbf{a}$, then $\mathbf{A}$ carries every vector in Π into a vector in Π, i.e., Π is an invariant plane of $\mathbf{A}$.

THEOREM 3. *The roots of the characteristic equation*

$$P(\lambda) = 0 \tag{1}$$

of a symmetric linear transformation $\mathbf{A}$ *are all real.*†

Proof. Suppose $\lambda = \alpha + i\beta$ is a complex root of (1). Then, since the coefficients of (1) are real, the number $\lambda^* = \alpha - i\beta$ conjugate to λ is

† Theorem 3 is also valid for arbitrary n, in particular for $n = 2$.

also a root of (1). Let $\mathbf{x}$ and $\mathbf{x}^*$ be eigenvectors corresponding to the eigenvalues λ and λ^*, so that

$$\mathbf{Ax} = \lambda\mathbf{x}, \qquad \mathbf{Ax}^* = \lambda^*\mathbf{x}^*.$$

Then, as noted in the remark on p. 110, $\mathbf{x}$ and $\mathbf{x}^*$ are complex conjugates, i.e.,

$$\mathbf{x} = x_k\mathbf{e}_k, \qquad \mathbf{x}^* = x_k^*\mathbf{e}_k,$$

where the numbers x_k and x_k^* are complex conjugates (as indicated by the notation). If $\lambda \neq \lambda^*$, it follows from Theorem 1 (which remains valid for complex eigenvalues) that the vectors $\mathbf{x}$ and $\mathbf{x}^*$ are orthogonal, so that $(\mathbf{x}, \mathbf{x}^*) = 0$. But, on the other hand,

$$(\mathbf{x}, \mathbf{x}^*) = x_k x_k^* = \sum_{k=1}^{3} |x_k|^2 > 0.$$

Therefore $\lambda = \lambda^*$, i.e., λ is real. ▮

THEOREM 4. *A symmetric linear transformation* **A** *of the space* L_3 *has three (pairwise) orthogonal eigenvectors.*

Proof. Let λ_1 be an eigenvalue of **A** (a real number, by Theorem 3), and let $\mathbf{a}_1$ be the corresponding eigenvector. Then, as noted in Remark 2, the plane Π orthogonal to $\mathbf{a}_1$ is invariant under **A**. The transformation **A** is again linear and symmetric in the plane Π. Let λ_2 be an eigenvalue of **A** and let $\mathbf{a}_2$ be the corresponding eigenvector (in Π). Then $\mathbf{a}_2$ is obviously orthogonal to $\mathbf{a}_1$. Now let $\mathbf{a}_3$ be a vector in Π which is orthogonal to $\mathbf{a}_2$. Then, by Theorem 2 (again see Remark 2), $\mathbf{a}_3$ is also an eigenvector of **A**. Thus we have found three (pairwise) orthogonal eigenvectors $\mathbf{a}_1, \mathbf{a}_2, \mathbf{a}_3$ of the transformation **A**. ▮

PROBLEMS

1. Prove Theorem 3 for the plane L_2 by direct evaluation of the roots of the characteristic equation of the transformation **A**.

2. Prove that if a linear transformation **A** of the space L_3 has three orthogonal eigenvectors, then **A** is symmetric.

3. Prove that two symmetric linear transformations of the space L_3 commute if and only if they share three orthogonal eigenvectors.

4. Let **A** be an antisymmetric linear transformation of the space L_3. Prove that
   a) Two (possibly complex) eigenvectors $\mathbf{a}_1$ and $\mathbf{a}_2$ of **A** corresponding to eigenvalues λ_1 and λ_2 such that $\lambda_1 + \lambda_2 \neq 0$ are orthogonal;
   b) If $\mathbf{a}$ is an eigenvector of **A**, then the plane orthogonal to $\mathbf{a}$ is invariant under **A**;
   c) The eigenvalues of **A** either vanish or are purely imaginary.

23. Diagonalization of a Symmetric Transformation

23.1. We begin with the following key

THEOREM. *Let* **A** *be a symmetric linear transformation of the space* L_3. *Then the matrix of* **A** *can always be reduced to diagonal form by transforming to a new orthonormal basis* $\mathbf{e}_{1'}, \mathbf{e}_{2'}, \mathbf{e}_{3'}$.

Proof. By Theorem 4 of the preceding section, **A** has three orthogonal eigenvectors $\mathbf{a}_1, \mathbf{a}_2, \mathbf{a}_3$. Suppose we normalize these vectors, by setting

$$\frac{\mathbf{a}_i}{|\mathbf{a}_i|} = \mathbf{e}_{i'}.$$

Then the vectors $\mathbf{e}_{1'}, \mathbf{e}_{2'}, \mathbf{e}_{3'}$ make up an orthonormal basis, and are also eigenvectors of **A**. Since

$$\mathbf{A}\mathbf{e}_{1'} = \lambda_1 \mathbf{e}_{1'}, \qquad \mathbf{A}\mathbf{e}_{2'} = \lambda_2 \mathbf{e}_{2'}, \qquad \mathbf{A}\mathbf{e}_{3'} = \lambda_3 \mathbf{e}_{3'},$$

the transformation **A** has the matrix

$$A' = \begin{pmatrix} \lambda_1 & 0 & 0 \\ 0 & \lambda_2 & 0 \\ 0 & 0 & \lambda_3 \end{pmatrix} \tag{1}$$

in this basis. Since the original basis $\mathbf{e}_1, \mathbf{e}_2, \mathbf{e}_3$ and the new basis $\mathbf{e}_{1'}, \mathbf{e}_{2'}, \mathbf{e}_{3'}$ are both orthonormal, the transformation

$$\mathbf{e}_{i'} = \gamma_{i'i} \mathbf{e}_i$$

from the former to the latter is described by an orthogonal matrix $\Gamma = (\gamma_{i'i})$, as in Sec. 6.1. We then have

$$A' = \Gamma A \Gamma^{-1},$$

where A is the matrix of **A** in the old basis and A' its matrix in the new basis (see Sec. 16.4). ▮

Remark. Geometrically the theorem means that a symmetric linear transformation is described by three simultaneous expansions (or contractions) along the three perpendicular axes determined by the vectors $\mathbf{e}_{1'}, \mathbf{e}_{2'}, \mathbf{e}_{3'}$, since a diagonal matrix like A' corresponds to just such a transformation (see Example 9, p. 73).

23.2. Next we examine the uniqueness of the basis $\mathbf{e}_{1'}, \mathbf{e}_{2'}, \mathbf{e}_{3'}$ in which the matrix of a symmetric linear transformation **A** takes the form (1). Here three cases arise:

Case 1. If the eigenvalues are distinct (so that $\lambda_1 \neq \lambda_2, \lambda_2 \neq \lambda_3, \lambda_1 \neq \lambda_3$) then the set of orthonormal eigenvectors $\mathbf{e}_{1'}, \mathbf{e}_{2'}, \mathbf{e}_{3'}$ is uniquely determined (to within reversal of directions and relabelling of vectors). In fact, if the vector **a** is not collinear with any of the vectors $\mathbf{e}_{1'}$, $\mathbf{e}_{2'}$, $\mathbf{e}_{3'}$, then **a** cannot be an eigenvector of **A**. For example, let

$$\mathbf{a} = \alpha\mathbf{e}_{1'} + \beta\mathbf{e}_{2'} \qquad (\alpha \neq 0, \quad \beta \neq 0).$$

Then the vector

$$\mathbf{A}\mathbf{a} = \alpha\lambda_1\mathbf{e}_{1'} + \beta\lambda_2\mathbf{e}_{2'}$$

is not collinear with **a**, since $\lambda_1 \neq \lambda_2$, and hence **a** is not an eigenvector of **A**.

Case 2. If two eigenvalues coincide (so that $\lambda_1 \neq \lambda_2, \lambda_2 = \lambda_3 = \lambda$, say) and if $\mathbf{e}_{1'}$, $\mathbf{e}_{2'}$, $\mathbf{e}_{3'}$ are the corresponding orthonormal eigenvectors, then every vector in the plane Π determined by $\mathbf{e}_{1'}$ and $\mathbf{e}_{2'}$ is an eigenvector of **A**. In fact, if

$$\mathbf{a} = \alpha\mathbf{e}_{2'} + \beta\mathbf{e}_{3'},$$

then

$$\mathbf{A}\mathbf{a} = \alpha\mathbf{A}\mathbf{e}_{2'} + \beta\mathbf{A}\mathbf{e}_{3'} = \alpha\lambda\mathbf{e}_{2'} + \beta\lambda\mathbf{e}_{3'} = \lambda(\alpha\mathbf{e}_{2'} + \beta\mathbf{e}_{3'}) = \lambda\mathbf{a}.$$

Therefore any orthogonal pair of vectors lying in Π can be chosen as the vectors $\mathbf{e}_{2'}$ and $\mathbf{e}_{3'}$. In this case, the transformation **A** represents the product of two transformations, a homothetic transformation with coefficient λ in the plane perpendicular to $\mathbf{e}_1$ and an expansion with coefficient λ_1 along the $\mathbf{e}_1$-axis.

Case 3. If all three eigenvalues coincide (so that $\lambda_1 = \lambda_2 = \lambda_3 = \lambda$), then every vector is an eigenvector (see Example 1, p. 108). Then **A** is a homothetic transformation with coefficient λ in the whole plane L_3, and any three orthonormal vectors can be chosen as the basis $\mathbf{e}_{1'}$, $\mathbf{e}_{2'}$, $\mathbf{e}_{3'}$.

Remark 1. The following observation facilitates the determination of the eigenvectors in Case 2: Since every vector in the plane Π is an eigenvector in this case, substitution of the eigenvalue $\lambda = \lambda_2 = \lambda_3$ into the system (3), p. 109 leads to the single equation

$$(a_{11} - \lambda)x_1 + a_{12}x_2 + a_{13}x_3 = 0 \tag{2}$$

(the other two equations of the system are proportional to this one). Every nontrivial solution of (2) determines an eigenvector corresponding to the eigenvalue $\lambda = \lambda_2 = \lambda_3$. Moreover, it follows from (2) that all the eigenvectors so obtained are perpendicular to the vector

$$\mathbf{a}_1 = (a_{11} - \lambda_2, a_{12}, a_{13}).$$

(Note that $\mathbf{a}_1 \neq \mathbf{0}$, since the coefficients of (2) cannot all vanish.) Hence $\mathbf{a}_1$ is an eigenvector corresponding to the eigenvalue λ_1. To construct the required orthonormal basis, we need only set

$$\mathbf{e}_{1'} = \frac{\mathbf{a}_1}{|\mathbf{a}_1|},$$

choose any normalized solution of (2) as the components of $\mathbf{e}_{2'}$, and then take $\mathbf{e}_{3'}$ to be the vector product $\mathbf{e}_{1'} \times \mathbf{e}_{2'}$.

Remark 2. For a symmetric linear transformation **A** of the plane L_2 there are only two possibilities:

Case 1. If the eigenvalues are distinct ($\lambda_1 \neq \lambda_2$), then **A** has the diagonal matrix

$$A = \begin{pmatrix} \lambda_1 & 0 \\ 0 & \lambda_2 \end{pmatrix}$$

in the basis consisting of the eigenvectors. Thus the transformation **A** is described by two simultaneous expansions (or contractions) along the pair of perpendicular axes determined by the eigenvectors $\mathbf{e}_{1'}$ and $\mathbf{e}_{2'}$ corresponding to λ_1 and λ_2.

Case 2. If the eigenvalues coincide ($\lambda_1 = \lambda_2 = \lambda$), then every vector in the plane L_2 is an eigenvector and **A** is a homothetic transformation in every orthonormal basis, with matrix

$$A = \begin{pmatrix} \lambda & 0 \\ 0 & \lambda \end{pmatrix}.$$

23.3. We now give some examples illustrating the above theory.

Example 1. Given a linear transformation **A** of the plane L_2 with matrix

$$A = \begin{pmatrix} 0 & 2 \\ 2 & 3 \end{pmatrix}$$

in an orthonormal basis $\mathbf{e}_1, \mathbf{e}_2$, find a new orthonormal basis in which the matrix of **A** is diagonal and write down the matrix.

Solution. The matrix of **A** is symmetric, and hence our problem is solvable. The characteristic equation of **A** is

$$\begin{vmatrix} -\lambda & 2 \\ 2 & 3 - \lambda \end{vmatrix} = 0$$

or

$$\lambda^2 - 3\lambda - 4 = 0,$$

with roots $\lambda_1 = 4$, $\lambda_2 = -1$. The next step is to find the eigenvectors corresponding to these eigenvalues.

1) For $\lambda = 4$ the system (3′), p. 112 takes the form

$$-4x_1 + 2x_2 = 0,$$

$$2x_1 - x_2 = 0,$$

with solution $x_1 = 1$, $x_2 = 2$ (say). Normalizing this solution, we get the unit eigenvector

$$\mathbf{e}_{1'} = \left(\frac{1}{\sqrt{5}}, \frac{2}{\sqrt{5}}\right)$$

corresponding to the eigenvalue $\lambda_1 = 4$.

2) For $\lambda = -1$ we get

$$x_1 + 2x_2 = 0,$$
$$2x_1 + 4x_2 = 0,$$

whence $x_1 = -2$, $x_2 = 1$ and

$$\mathbf{e}_{2'} = \left(-\frac{2}{\sqrt{5}}, \frac{1}{\sqrt{5}}\right).$$

In transforming to the basis $\mathbf{e}_{1'}, \mathbf{e}_{2'}$, the components of all vectors transform according to the formula

$$x_{i'} = \gamma_{i'i} x_i$$

(see Sec. 6.2), where

$$\Gamma = (\gamma_{i'i}) = \begin{pmatrix} \frac{1}{\sqrt{5}} & \frac{2}{\sqrt{5}} \\ -\frac{2}{\sqrt{5}} & \frac{1}{\sqrt{5}} \end{pmatrix}.$$

In the basis $\mathbf{e}_{1'}, \mathbf{e}_{2'}$ the matrix of the transformation $\mathbf{A}$ takes the form

$$A' = \Gamma A \Gamma^{-1} = \begin{pmatrix} \frac{1}{\sqrt{5}} & \frac{2}{\sqrt{5}} \\ -\frac{2}{\sqrt{5}} & \frac{1}{\sqrt{5}} \end{pmatrix} \begin{pmatrix} 0 & 2 \\ 2 & 3 \end{pmatrix} \begin{pmatrix} \frac{1}{\sqrt{5}} & -\frac{2}{\sqrt{5}} \\ \frac{2}{\sqrt{5}} & \frac{1}{\sqrt{5}} \end{pmatrix} = \begin{pmatrix} 4 & 0 \\ 0 & -1 \end{pmatrix}$$

Note that the matrix A' can be written down without carrying out these calculations, since its diagonal elements are just the eigenvalues of A. The transformation $\mathbf{A}$ corresponds to an expansion along $\mathbf{e}_{1'}$ with coefficient 4, together with an expansion along $\mathbf{e}_{2'}$ with coefficient -1 (actually a reflection in the line of $\mathbf{e}_{1'}$).

Example 2. Given a linear transformation $\mathbf{A}$ of the space L_3 with matrix

$$A = \begin{pmatrix} 1 & 1 & 3 \\ 1 & 5 & 1 \\ 3 & 1 & 1 \end{pmatrix}$$

in an orthonormal basis $\mathbf{e}_1, \mathbf{e}_2, \mathbf{e}_3$, find a new orthonormal basis in which the matrix of $\mathbf{A}$ is diagonal and write down the matrix.

Solution. The problem can be solved since A is symmetric. The characteristic equation is

$$\begin{vmatrix} 1-\lambda & 1 & 3 \\ 1 & 5-\lambda & 1 \\ 3 & 1 & 1-\lambda \end{vmatrix} = 0$$

or

$$\lambda^3 - 7\lambda^2 + 36 = 0,$$

with roots $\lambda_1 = 6$, $\lambda_2 = 3$, $\lambda_3 = -2$. Since these roots are distinct, the transformation **A** is of the type described in Case 1, p. 127. Here the system (3), p. 109 takes the form

$$\begin{aligned}(1-\lambda)x_1 + x_2 + 3x_3 &= 0,\\ x_1 + (5-\lambda)x_2 + x_3 &= 0,\\ 3x_1 + x_2 + (1-\lambda)x_3 &= 0.\end{aligned}$$

Substituting $\lambda = 6$, $\lambda = 3$, $\lambda = -2$ in turn into this system, we get the vectors of the new orthonormal basis:

$$\mathbf{e}_{1'} = \left(\frac{1}{\sqrt{6}}, \frac{2}{\sqrt{6}}, \frac{1}{\sqrt{6}}\right),$$

$$\mathbf{e}_{2'} = \left(\frac{1}{\sqrt{3}}, -\frac{1}{\sqrt{3}}, \frac{1}{\sqrt{3}}\right),$$

$$\mathbf{e}_{3'} = \left(\frac{1}{\sqrt{2}}, 0, -\frac{1}{\sqrt{2}}\right).$$

Note that the vector $\mathbf{e}_{3'}$ is just the vector product $\mathbf{e}_{1'} \times \mathbf{e}_{2'}$. Here the matrix Γ is given by

$$\Gamma = \begin{pmatrix} \frac{1}{\sqrt{6}} & \frac{2}{\sqrt{6}} & \frac{1}{\sqrt{6}} \\ \frac{1}{\sqrt{3}} & -\frac{1}{\sqrt{3}} & \frac{1}{\sqrt{3}} \\ \frac{1}{\sqrt{2}} & 0 & -\frac{1}{\sqrt{2}} \end{pmatrix},$$

while **A** has the matrix

$$A' = \Gamma A \Gamma^{-1} = \begin{pmatrix} 6 & 0 & 0 \\ 0 & 3 & 0 \\ 0 & 0 & -2 \end{pmatrix}$$

in the basis $\mathbf{e}_{1'}, \mathbf{e}_{2'}, \mathbf{e}_{3'}$. Geometrically, the transformation **A** represents three simultaneous expansions along the axes $\mathbf{e}_{1'}$, $\mathbf{e}_{2'}$ and $\mathbf{e}_{3'}$ with coefficients 6, 3 and -2, respectively.

Example 3. Given a linear transformation **A** of the space L_3 with matrix

$$A = \begin{pmatrix} 5 & 2 & 2 \\ 2 & 2 & -4 \\ 2 & -4 & 2 \end{pmatrix}$$

in an orthonormal basis $\mathbf{e}_1, \mathbf{e}_2, \mathbf{e}_3$, find a new orthonormal basis in which the matrix of **A** is diagonal and write down the matrix.

Solution. Once again the solvability of the problem is guaranteed by the symmetry of **A**. This time the characteristic equation is

$$\begin{vmatrix} 5-\lambda & 2 & 2 \\ 2 & 2-\lambda & -4 \\ 2 & -4 & 2-\lambda \end{vmatrix} = 0$$

or

$$\lambda^3 - 9\lambda^2 + 108 = 0,$$

with roots $\lambda_1 = -3$, $\lambda_2 = \lambda_3 = 6$. Thus we are now dealing with Case 2, p. 128. According to Remark 1, p. 128, the system corresponding to the eigenvalue $\lambda_2 = \lambda_3 = 6$ reduces to the single equation

$$-x_1 + 2x_2 + 3x_3 = 0. \qquad (3)$$

It follows that $\mathbf{a}_1 = (-1, 2, 3)$ is the eigenvector corresponding to the eigenvalue $\lambda_1 = -3$. The corresponding unit vector is just

$$\mathbf{e}_{1'} = (-\tfrac{1}{3}, \tfrac{2}{3}, \tfrac{2}{3}).$$

We now take any solution of (3), say $x_1 = 2$, $x_2 = -1$, $x_3 = 2$, and normalize it, obtaining the eigenvector

$$\mathbf{e}_{2'} = (\tfrac{2}{3}, -\tfrac{1}{3}, \tfrac{2}{3}).$$

Finally the eigenvector $\mathbf{e}_{3'}$ is given by the vector product

$$\mathbf{e}_{3'} = \mathbf{e}_{1'} \times \mathbf{e}_{2'} = (\tfrac{2}{3}, \tfrac{2}{3}, -\tfrac{1}{3}).$$

Hence the matrix of the transformation to the new basis is

$$\Gamma = \begin{pmatrix} -\frac{1}{3} & \frac{2}{3} & \frac{2}{3} \\ \frac{2}{3} & -\frac{1}{3} & \frac{2}{3} \\ \frac{2}{3} & \frac{2}{3} & -\frac{1}{3} \end{pmatrix},$$

while A has the matrix

$$A' = \Gamma A \Gamma^{-1} = \begin{pmatrix} -3 & 0 & 0 \\ 0 & 6 & 0 \\ 0 & 0 & 6 \end{pmatrix}$$

in the basis $\mathbf{e}_{1'}, \mathbf{e}_{2'}, \mathbf{e}_{3'}$. Geometrically, the transformation **A** represents an expansion along the $\mathbf{e}_{1'}$-axis with coefficient -3, together with a homothetic transformation in the $\mathbf{e}_{2'}, \mathbf{e}_{3'}$-plane with coefficient 6.

PROBLEMS

1. Given a symmetric linear transformation **A** of the plane L_2 or of the space L_3 with each of the following matrices in some orthonormal basis, find a new orthonormal basis in which the matrix of **A** takes diagonal form and write down the matrix:

a) $\begin{pmatrix} 6 & 2 \\ 2 & 3 \end{pmatrix}$; b) $\begin{pmatrix} 7 & -2 & 0 \\ -2 & 6 & -2 \\ 0 & -2 & 5 \end{pmatrix}$; c) $\begin{pmatrix} 1 & 2 & -4 \\ 2 & -2 & -2 \\ -4 & -2 & 1 \end{pmatrix}$;

d) $\begin{pmatrix} 0 & 0 & 1 \\ 0 & 1 & 0 \\ 1 & 0 & 0 \end{pmatrix}$.

2. Raise the following matrices to the thirtieth power:

a) $\begin{pmatrix} 6 & 2 \\ 2 & 3 \end{pmatrix}$; b) $\begin{pmatrix} 7 & -2 & 0 \\ -2 & 6 & -2 \\ 0 & -2 & 5 \end{pmatrix}$.

3. A symmetric linear transformation **A** is called *nonnegative* if $(\mathbf{x}, \mathbf{Ax}) \geq 0$ for every vector **x**. Prove that

a) **A** is nonnegative if and only if all its eigenvalues are nonnegative;

b) If **A** is nonnegative, there is a nonnegative symmetric transformation **B** such that $\mathbf{B}^2 = \mathbf{A}$;

c) If **A** is nonnegative, and if $\mathbf{AC} = \mathbf{CA}$ for some transformation **C**, then $\mathbf{BC} = \mathbf{CB}$ where $\mathbf{B}^2 = \mathbf{A}$;

d) The sum of two nonnegative transformations is nonnegative;

e) The product of two commuting nonnegative transformations is nonnegative;

f) If $P(\lambda)$ is a polynomial with nonnegative coefficients and if **A** is nonnegative, then $P(\mathbf{A})$ is nonnegative;

g) **A** is nonnegative if and only if the coefficients of the characteristic polynomial of **A** alternate in sign.

4. Prove that the symmetric transformations with the following matrices in some orthonormal basis are nonnegative:

a) $A = \begin{pmatrix} 4 & 2 & 4 \\ 2 & 1 & 2 \\ 4 & 2 & 4 \end{pmatrix}$; b) $A = \begin{pmatrix} 13 & 14 & 4 \\ 14 & 24 & 18 \\ 4 & 18 & 29 \end{pmatrix}$.

In each case, find the matrix (in the same basis) of the transformation **B** such that $\mathbf{B}^2 = \mathbf{A}$.

5. Prove that if **A** is a symmetric orthogonal transformation, then its matrix can be reduced to one of the following four forms by making an orthogonal transformation:

$$\begin{pmatrix} 1 & 0 & 0 \\ 0 & 1 & 0 \\ 0 & 0 & 1 \end{pmatrix}, \quad \begin{pmatrix} 1 & 0 & 0 \\ 0 & 1 & 0 \\ 0 & 0 & -1 \end{pmatrix}, \quad \begin{pmatrix} 1 & 0 & 0 \\ 0 & -1 & 0 \\ 0 & 0 & -1 \end{pmatrix}, \quad \begin{pmatrix} -1 & 0 & 0 \\ 0 & -1 & 0 \\ 0 & 0 & -1 \end{pmatrix}.$$

24. Reduction of a Quadratic Form to Canonical Form

As shown in Sec. 15.4, there is a one-to-one correspondence between symmetric linear transformations and quadratic forms. Using this correspondence, together with the fact that the matrix of a symmetric linear transformation can be reduced to diagonal form, we are now in a position to make a related simplification of quadratic forms:

THEOREM.† *Let* **A** *be a symmetric linear transformation of the space* L_3, *and let*

$$\varphi = (\mathbf{x}, \mathbf{Ax}) = a_{ij}x_ix_j$$

be the corresponding quadratic form, where $(a_{ij}) = A$ *is the matrix of* **A**. *Then* φ *takes the "canonical form"*

$$\varphi = \lambda_1 x_{1'}^2 + \lambda_2 x_{2'}^2 + \lambda_3 x_{3'}^2 \qquad (1)$$

in the orthonormal basis $\mathbf{e}_{1'}, \mathbf{e}_{2'}, \mathbf{e}_{3'}$‡ *in which* A *takes the diagonal form*

$$\begin{pmatrix} \lambda_1 & 0 & 0 \\ 0 & \lambda_2 & 0 \\ 0 & 0 & \lambda_3 \end{pmatrix}$$

(involving the eigenvalues $\lambda_1, \lambda_2, \lambda_3$ *of* **A**).

Proof. Let $\mathbf{x} = x_{i'}\mathbf{e}_{i'}$, where the numbers $x_{i'}$ are the components of the vector **x** with respect to the basis $\mathbf{e}_{1'}, \mathbf{e}_{2'}, \mathbf{e}_{3'}$. Then, since $\mathbf{e}_{i'}$ is an eigenvector of **A** with eigenvalue λ_i, we have

$$\begin{aligned} \varphi = (\mathbf{x}, \mathbf{Ax}) &= (x_{i'}\mathbf{e}_{i'}, \mathbf{A}x_{j'}\mathbf{e}_{j'}) = (x_{i'}\mathbf{e}_{i}, x_{j'}\lambda_j\mathbf{e}_{j'}) \\ &= \delta_{ij}\lambda_j x_{i'}x_{j'} = \lambda_i x_{i'}^2 = \lambda_1 x_{1'}^2 + \lambda_2 x_{2'}^2 + \lambda_3 x_{3'}^2. \quad \blacksquare \end{aligned}$$

Remark. The directions of $\mathbf{e}_{1'}, \mathbf{e}_{2'}, \mathbf{e}_{3'}$ are called the *principal directions* of the form φ corresponding to the eigenvalues $\lambda_1, \lambda_2, \lambda_3$. It follows from the results of Sec. 23.2 that if $\lambda_1 \neq \lambda_2, \lambda_2 \neq \lambda_3, \lambda_3 \neq \lambda_1$, then φ has exactly three principal directions, if $\lambda_1 \neq \lambda_2 = \lambda_3$, then φ has one principal direction corresponding to λ_1 and infinitely many principal directions perpendicular to the direction corresponding to λ_1, while if $\lambda_1 = \lambda_2 = \lambda_3$, then every direction in space is a principal direction of φ.

Example 1. Reduce

$$\varphi = 4x_1x_2 + 3x_2^2$$

to canonical form.

Solution. The symmetric linear transformation **A** corresponding to φ has the matrix

$$A = \begin{pmatrix} 0 & 2 \\ 2 & 3 \end{pmatrix}.$$

This is just the matrix of the transformation considered in Example 1, p. 129, with eigenvalues $\lambda_1 = 4$, $\lambda_2 = -1$. Hence we can reduce φ to the sum

† The theorem has an obvious analogue for a symmetric linear transformation of the plane L_2.

‡ The existence of such a basis is guaranteed by the theorem on p. 127.

of squares

$$\varphi = 4x_{1'}^2 - x_{2'}^2$$

by going over to the basis $\mathbf{e}_1, \mathbf{e}_{2'}$ found on p. 130.

Example 2. Reduce

$$\varphi = x_1^2 + 5x_2^2 + x_3^2 + 2x_1x_2 + 6x_1x_3 + 2x_2x_3$$

to canonical form.

Solution. The matrix of the symmetric linear transformation $\mathbf{A}$ corresponding to φ is

$$A = \begin{pmatrix} 1 & 1 & 3 \\ 1 & 5 & 1 \\ 3 & 1 & 1 \end{pmatrix},$$

and coincides with the matrix of the transformation considered in Example 2, p. 130, with eigenvalues $\lambda_1 = 6$, $\lambda_2 = 3$, $\lambda_3 = -2$. Hence we can reduce φ to the sum of squares

$$\varphi = 6x_{1'}^2 + 3x_{2'}^2 - 2x_{3'}^2$$

by going over to the basis $\mathbf{e}_{1'}, \mathbf{e}_{2'}, \mathbf{e}_{3'}$ found on p. 131.

24.2. A quadratic form $\varphi(\mathbf{x}, \mathbf{x})$ is called *positive* (or *negative*) *definite* if it takes only positive (or negative) values for every vector $\mathbf{x} \neq \mathbf{0}$. Since this must be true in any basis, it must hold in particular in the basis $\mathbf{e}_{1'}, \mathbf{e}_{2'}, \mathbf{e}_{3'}$ in which φ has the canonical form (1). Clearly, the expression (1) is positive (or negative) for arbitrary $x_{1'}, x_{2'}, x_{3'}$ if and only if the eigenvalues $\lambda_1, \lambda_2, \lambda_3$ are positive (or negative). However, it is important to have a condition allowing us to determine whether or not a given quadratic form $\varphi(\mathbf{x}, \mathbf{x})$ is positive or negative definite when it is specified in an arbitrary orthonormal basis $\mathbf{e}_1, \mathbf{e}_2, \mathbf{e}_3$ (not necessarily the "canonical basis" $\mathbf{e}_{1'}, \mathbf{e}_{2'}, \mathbf{e}_{3'}$). Let $A = (a_{ij})$ be the matrix of $\varphi(\mathbf{x}, \mathbf{x})$ in the basis $\mathbf{e}_1, \mathbf{e}_2, \mathbf{e}_3$. Then the quantities

$$M_1 = a_{11}, \qquad M_2 = \begin{vmatrix} a_{11} & a_{12} \\ a_{21} & a_{22} \end{vmatrix}, \qquad M_3 = \begin{vmatrix} a_{11} & a_{12} & a_{13} \\ a_{21} & a_{22} & a_{23} \\ a_{31} & a_{32} & a_{33} \end{vmatrix}$$

are called the (*descending*) *principal minors* of A. The desired condition for $\varphi(x, x)$ to be positive definite is given by the following

THEOREM (***Sylvester's criterion***). *A quadratic form*

$$\varphi(\mathbf{x}, \mathbf{x}) = a_{ij}x_ix_j$$

with matrix $A = (a_{ij})$ is positive definite if and only if its principal minors (in any given basis) are positive.

Proof. First we prove the theorem for the case of a quadratic form

$$\varphi(\mathbf{x}, \mathbf{x}) = a_{11}x_1^2 + 2a_{12}x_1x_2 + a_{22}x_2^2 \tag{2}$$

defined in the plane L_2. In terms of a new auxiliary variable $t = x_1/x_2$, we can write (2) as

$$\varphi(\mathbf{x}, \mathbf{x}) = x_2^2(a_{11}t^2 + 2a_{12}t + a_{22}).$$

The principal minor M_2 of the form $\varphi(\mathbf{x}, \mathbf{x})$ differs only in sign from the discriminant $D = a_{12}^2 - a_{11}a_{22}$ of the quadratic polynomial in parentheses. If $M_2 > 0$, then $D < 0$ and the polynomial does not change sign as the parameter t varies. If $M_1 = a_{11} > 0$, this sign will be positive for all t (why?). Hence $\varphi(\mathbf{x}, \mathbf{x})$ is positive definite in the plane L_2 if $M_1 > 0$, $M_2 > 0$. Conversely, it is easy to see that $\varphi(\mathbf{x}, \mathbf{x}) > 0$ implies $M_1 > 0$, $M_2 > 0$. In fact, if

$$\mathbf{x}_1 = \mathbf{e}_1, \qquad \mathbf{x}_2 = -a_{12}\mathbf{e}_1 + a_{11}\mathbf{e}_2,$$

then

$$\varphi(\mathbf{x}_1, \mathbf{x}_1) = a_{11} = M_1,$$
$$\varphi(\mathbf{x}_2, \mathbf{x}_2) = a_{11}(a_{11}a_{22} - a_{12}^2) = M_1M_2,$$

which implies $M_1 > 0$, $M_2 > 0$ since $\varphi(\mathbf{x}_1, \mathbf{x}_1) > 0$, $\varphi(\mathbf{x}_2, \mathbf{x}_2) > 0$.

Turning now to the proof of Sylvester's criterion in three dimensions, we note that

$$\varphi(\mathbf{x}, \mathbf{x}) = a_{11}x_1^2 + a_{22}x_2^2 + a_{33}x_3^2 + 2a_{12}x_1x_2 + 2a_{13}x_1x_3 + 2a_{23}x_2x_3$$

in the given basis $\mathbf{e}_1, \mathbf{e}_2, \mathbf{e}_3$, which becomes

$$\varphi(\mathbf{x}, \mathbf{x}) = \lambda_1x_{1'}^2 + \lambda_2x_{2'}^2 + \lambda_3x_{3'}^2$$

after going over to the basis $\mathbf{e}_{1'}, \mathbf{e}_{2'}, \mathbf{e}_{3'}$ made up of the vectors directed along the principal directions of $\varphi(\mathbf{x}, \mathbf{x})$. Since the principal minor M_3 coincides with the invariant I_3 of $\varphi(\mathbf{x}, \mathbf{x})$, we have $M_3 = \lambda_1\lambda_2\lambda_3$ (see Prob. 6, p. 115). Suppose $\varphi(\mathbf{x}, \mathbf{x})$ is positive definite. Then $\lambda_1 > 0$, $\lambda_2 > 0$, $\lambda_3 > 0$, and hence $M_3 > 0$. To prove that M_1 and M_2 are also positive in this case, we need only consider the form $\varphi(\mathbf{x}, \mathbf{x})$ in the plane $x_3 = 0$ and use Sylvester's criterion in two dimensions (just proved above). Conversely, suppose all three principal minors of the form $\varphi(\mathbf{x}, \mathbf{x})$ are positive. Then

$$M_3 = \lambda_1\lambda_2\lambda_3 > 0,$$

and there are just two possibilities: 1) All three eigenvalues are positive, or 2) One of the eigenvalues is positive and the other two are negative. In the first case, the quadratic form $\varphi(\mathbf{x}, \mathbf{x})$ is positive definite and our converse assertion is proved. Thus suppose one of the numbers λ_i is positive, say $\lambda_2 > 0$, while the other two are negative, say $\lambda_1 < 0$, $\lambda_3 < 0$. Then the form $\varphi(\mathbf{x}, \mathbf{x})$ is negative definite in the $\mathbf{e}_{1'}, \mathbf{e}_{3'}$-plane. But, on the other hand, the form $\varphi(x, x)$ reduces to

$$\varphi(\mathbf{x}, \mathbf{x}) = a_{11}x_1^2 + 2a_{12}x_1x_2 + a_{22}x_2^2$$

in the $\mathbf{e}_1, \mathbf{e}_2$-plane, and is positive definite in this plane because of the positivity of the first two principal minors. It follows that $\varphi(\mathbf{x}, \mathbf{x})$ is simultaneously positive definite and negative definite on the line of intersection of the $\mathbf{e}_1, \mathbf{e}_2$-plane and the $\mathbf{e}_{1'}, \mathbf{e}_{3'}$-plane. This contradiction shows that the numbers λ_i must all be positive. ∎

Remark. The quadratic form

$$\varphi(\mathbf{x}, \mathbf{x}) = a_{ij}x_i x_j \tag{3}$$

is positive definite if and only if the form

$$-\varphi(\mathbf{x}, \mathbf{x}) = -a_{ij}x_i x_j$$

is negative definite. Hence (3) is negative definite if and only if

$$a_{11} < 0, \qquad \begin{vmatrix} a_{11} & a_{12} \\ a_{21} & a_{22} \end{vmatrix} > 0, \qquad \begin{vmatrix} a_{11} & a_{12} & a_{13} \\ a_{21} & a_{22} & a_{23} \\ a_{31} & a_{32} & a_{33} \end{vmatrix} < 0.$$

PROBLEMS

1. Find the canonical form to which each of the following quadratic forms can be reduced by an orthogonal transformation without carrying out the transformation explicitly:

a) $\varphi = x_1x_2$;
b) $\varphi = x_1^2 + 2x_1x_2 + x_2^2$;
c) $\varphi = x_1^2 + x_1x_2 + x_2^2$;
d) $\varphi = 3x_2^2 + 3x_3^2 + 4x_1x_2 + 4x_1x_3 - 2x_2x_3$;
e) $\varphi = x_1^2 - 2x_1x_2 - 2x_1x_3 - 2x_2x_3$.

2. Find an orthogonal transformation Γ reducing each of the following quadratic forms to canonical form:

a) $\varphi = 5x_1^2 + 8x_1x_2 + 5x_2^2$;
b) $\varphi = x_1x_2 + x_2x_3 + x_3x_1$;
c) $\varphi = 7x_1^2 + 6x_2^2 + 5x_3^2 - 4x_1x_2 - 4x_2x_3$;
d) $\varphi = 2x_1^2 + x_2^2 - 4x_1x_2 - 4x_2x_3$;
e) $\varphi = 3x_1^2 + 6x_2^2 + 3x_3^2 - 4x_1x_2 - 8x_1x_3 - 4x_2x_3$.

3. For which values of the parameter a is each of the following quadratic forms positive definite:

a) $\varphi = 3x_1^2 - 4x_1x_2 + 4ax_2^2$;
b) $\varphi = 5x_1^2 + x_2^2 + ax_3^2 + 4x_1x_2 - 2x_1x_3 - 2x_2x_3$;
c) $\varphi = 2x_1^2 + x_2^2 + 3x_3^2 + 2ax_1x_2 + 2x_1x_3$?

4. Let λ_1 and λ_2 be the eigenvalues of the symmetric linear transformation corresponding to a quadratic form $\varphi(\mathbf{x}, \mathbf{x})$ defined in the plane L_2. Prove that if $\lambda_1 \le \lambda_2$, then

$$\lambda_1(\mathbf{x}, \mathbf{x}) \le \varphi(\mathbf{x}, \mathbf{x}) \le \lambda_2(\mathbf{x}, \mathbf{x}).$$

5. Prove that the eigenvalues of a symmetric matrix A all lie in the interval $[a, b]$ if and only if the quadratic form with matrix $A - xE$ is positive definite for all $x < a$ and negative definite for all $x > b$.

6. Let $\varphi(\mathbf{x}, \mathbf{x}) = 1$ be the characteristic surface of a symmetric linear transformation **A** (see Sec. 12.5). Determine the form of the surface if the eigenvalues of **A** satisfy the following conditions:

a) $\lambda_1 = \lambda_2 > 0, \lambda_3 < 0$;
b) $\lambda_1 = \lambda_2 < 0, \lambda_3 > 0$;
c) $\lambda_1 > 0, \lambda_2 > 0, \lambda_3 > 0$;
d) $\lambda_1 > 0, \lambda_2 > 0, \lambda_3 < 0$;
e) $\lambda_1 > 0, \lambda_2 < 0, \lambda_3 < 0$;
f) $\lambda_1 < 0, \lambda_2 < 0, \lambda_3 < 0$.

25. Representation of a Nonsingular Transformation

25.1. As the following remarkable theorem shows, orthogonal transformations and symmetric transformations suffice, in a certain sense, to describe arbitrary linear transformations:

THEOREM.† *Every nonsingular linear transformation* **A** *of the space* L_3 *can be represented as the product of an orthogonal transformation and a symmetric linear transformation.*

Proof. Let **A*** be the adjoint of **A**. Then the transformation **A*A** is symmetric, since

$$(\mathbf{A}^*\mathbf{A})^* = \mathbf{A}^*(\mathbf{A}^*)^* = \mathbf{A}^*\mathbf{A}$$

by the theorem on p. 89. Being symmetric, the transformation **A*A** has three orthonormal eigenvectors $\mathbf{e}_{1'}, \mathbf{e}_{2'}, \mathbf{e}_{3'}$, by Theorem 4, p. 126, so that

$$(\mathbf{A}^*\mathbf{A})\mathbf{e}_{1'} = \lambda_1\mathbf{e}_{1'}, \qquad (\mathbf{A}^*\mathbf{A})\mathbf{e}_{2'} = \lambda_2\mathbf{e}_{2'}, \qquad (\mathbf{A}^*\mathbf{A})\mathbf{e}_{3'} = \lambda_3\mathbf{e}_{3'}. \qquad (1)$$

Moreover, the eigenvalues $\lambda_1, \lambda_2, \lambda_3$ of the transformation **A*A** are all nonnegative. In fact, it follows from (1) that

$$\lambda_i = (\mathbf{e}_{i'}, (\mathbf{A}^*\mathbf{A})\mathbf{e}_{i'}) = (\mathbf{e}_{i'}, \mathbf{A}^*(\mathbf{A}\mathbf{e}_{i'})) = (\mathbf{A}\mathbf{e}_{i'}, \mathbf{A}\mathbf{e}_{i'}) \geqslant 0$$

(cf. Prob. 8, p. 93).

Now let **H** be the transformation with matrix

$$H' = \begin{pmatrix} \sqrt{\lambda_1} & 0 & 0 \\ 0 & \sqrt{\lambda_2} & 0 \\ 0 & 0 & \sqrt{\lambda_3} \end{pmatrix}$$

in the basis $\mathbf{e}_{1'}, \mathbf{e}_{2'}, \mathbf{e}_{3'}$. Since H' is a symmetric matrix, **H** is a symmetric linear transformation (why?). Moreover, the transformation $\mathbf{H}^2$ has the matrix

$$H'^2 = \begin{pmatrix} \lambda_1 & 0 & 0 \\ 0 & \lambda_2 & 0 \\ 0 & 0 & \lambda_3 \end{pmatrix}$$

† The theorem has an obvious analogue for a nonsingular linear transformation of the plane L_2.

in the basis $\mathbf{e}_{1'}, \mathbf{e}_{2'}, \mathbf{e}_{3'}$, i.e., the same matrix as the transformation $\mathbf{A}^*\mathbf{A}$. It follows that

$$\mathbf{A}^*\mathbf{A} = \mathbf{H}^2,$$

and hence

$$\mathbf{A} = (\mathbf{A}^*)^{-1}\mathbf{H}^2 = ((\mathbf{A}^*)^{-1}\mathbf{H})\mathbf{H},$$

and the theorem will be proved if we succeed in showing that the transformation

$$\mathbf{T} = (\mathbf{A}^*)^{-1}\mathbf{H} \tag{2}$$

is orthogonal. But

$$\mathbf{T}^* = ((\mathbf{A}^*)^{-1}\mathbf{H})^* = \mathbf{H}^*((\mathbf{A}^*)^{-1})^* = \mathbf{H}\mathbf{A}^{-1},$$

where we use the symmetry of $\mathbf{H}$ and the fact that $((\mathbf{A}^*)^{-1})^* = \mathbf{A}^{-1}$. It follows that

$$\mathbf{T}\mathbf{T}^* = (\mathbf{A}^*)^{-1}\mathbf{H}\mathbf{H}\mathbf{A}^{-1} = (\mathbf{A}^*)^{-1}\mathbf{H}^2\mathbf{A}^{-1} = (\mathbf{A}^*)^{-1}\mathbf{A}^*\mathbf{A}\mathbf{A}^{-1} = \mathbf{E}\mathbf{E} = \mathbf{E},$$

i.e., $\mathbf{T}$ is in fact an orthogonal transformation. Thus, finally,

$$\mathbf{A} = \mathbf{T}\mathbf{H}, \tag{3}$$

where $\mathbf{T}$ is orthogonal and $\mathbf{H}$ symmetric. ▮

Remark 1. Geometrically, the theorem means that any nonsingular linear transformation consists of three simultaneous expansions (or contractions) along three perpendicular axes, followed by a rotation of the whole space (together with these axes) about the origin.†

Remark 2. In proving the theorem, we have given an explicit procedure for constructing the symmetric and orthogonal transformations figuring in the representation (3). Note that to find the matrix of the orthogonal transformation (2), we must first find the matrix of the symmetric transformation $\mathbf{H}$ in the original basis $\mathbf{e}_1, \mathbf{e}_2, \mathbf{e}_3$ by using the formula

$$H = \Gamma^{-1}H'\Gamma,$$

where Γ is the matrix of the (orthogonal) transformation from the basis $\mathbf{e}_1, \mathbf{e}_2, \mathbf{e}_3$ to the basis $\mathbf{e}_{1'}, \mathbf{e}_{2'}, \mathbf{e}_{3'}$.

25.2. We now give two examples illustrating the above theory.

Example 1. Let $\mathbf{A}$ be the linear transformation of the plane L_2 with matrix

$$A = \begin{pmatrix} -\frac{36}{25} & \frac{2}{25} \\ -\frac{23}{25} & \frac{36}{25} \end{pmatrix}$$

in some orthonormal basis $\mathbf{e}_1, \mathbf{e}_2$. Express $\mathbf{A}$ as a product of an orthogonal transformation and a symmetric transformation.

† See Prob. 7, p. 121. If $\mathbf{T}$ is improper, the rotation is coupled with a reflection.

Solution. First we find the symmetric transformation **A*A** and reduce it to diagonal form. This transformation has the matrix

$$A^*A = \begin{pmatrix} -\frac{36}{25} & -\frac{23}{25} \\ \frac{2}{25} & \frac{36}{25} \end{pmatrix} \begin{pmatrix} -\frac{36}{25} & \frac{2}{25} \\ -\frac{23}{25} & \frac{36}{25} \end{pmatrix} = \begin{pmatrix} \frac{73}{25} & -\frac{36}{25} \\ -\frac{36}{25} & \frac{52}{25} \end{pmatrix}$$

and characteristic equation

$$\begin{vmatrix} \frac{73}{25} - \lambda & -\frac{36}{25} \\ -\frac{36}{25} & \frac{52}{25} - \lambda \end{vmatrix} = 0,$$

which simplifies to

$$\lambda^2 - 5\lambda + 4 = 0.$$

Hence the eigenvalues of **A*A** are $\lambda_1 = 1$, $\lambda_2 = 4$, with corresponding orthonormal eigenvectors

$$\mathbf{e}_{1'} = \left(\frac{3}{5}, \frac{4}{5}\right), \qquad \mathbf{e}_{2'} = \left(\frac{4}{5}, -\frac{3}{5}\right),$$

and the matrix of the transformation **A*A** takes the form

$$H^2 = \begin{pmatrix} 1 & 0 \\ 0 & 4 \end{pmatrix}$$

in the basis $\mathbf{e}_{1'}, \mathbf{e}_{2'}$. The required symmetric transformation **H** has the matrix

$$H' = \begin{pmatrix} 1 & 0 \\ 0 & 2 \end{pmatrix}$$

in the basis $\mathbf{e}_{1'}, \mathbf{e}_{2'}$, and the matrix

$$H = \Gamma^{-1}H'\Gamma = \begin{pmatrix} \frac{3}{5} & \frac{4}{5} \\ \frac{4}{5} & -\frac{3}{5} \end{pmatrix} \begin{pmatrix} 1 & 0 \\ 0 & 2 \end{pmatrix} \begin{pmatrix} \frac{3}{5} & \frac{4}{5} \\ \frac{4}{5} & -\frac{3}{5} \end{pmatrix} = \begin{pmatrix} \frac{41}{25} & -\frac{12}{25} \\ -\frac{12}{25} & \frac{34}{25} \end{pmatrix}$$

in the original basis $\mathbf{e}_1, \mathbf{e}_2$.

We can now construct the matrix of the orthogonal transformation $\mathbf{T} = (\mathbf{A}^*)^{-1}\mathbf{H}$. Observing first that

$$A^* = \begin{pmatrix} -\frac{36}{25} & -\frac{23}{25} \\ \frac{2}{25} & \frac{36}{25} \end{pmatrix}, \qquad (A^*)^{-1} = \begin{pmatrix} -\frac{18}{25} & -\frac{23}{50} \\ \frac{1}{25} & \frac{18}{25} \end{pmatrix},$$

we have

$$T = (A^*)^{-1}H = \begin{pmatrix} -\frac{18}{25} & -\frac{23}{50} \\ \frac{1}{25} & \frac{18}{25} \end{pmatrix} \begin{pmatrix} \frac{41}{25} & -\frac{12}{25} \\ -\frac{12}{25} & \frac{34}{25} \end{pmatrix} = \begin{pmatrix} -\frac{24}{25} & -\frac{7}{25} \\ -\frac{7}{25} & \frac{24}{25} \end{pmatrix}.$$

Noting that

$$T = \begin{pmatrix} \frac{24}{25} & -\frac{7}{25} \\ \frac{7}{25} & \frac{24}{25} \end{pmatrix} \begin{pmatrix} -1 & 0 \\ 0 & 1 \end{pmatrix},$$

we see that the transformation **A** consists of two simultaneous expansions along the $\mathbf{e}_{1'}$ - and $\mathbf{e}_{2'}$-axes, with coefficients 1 and 2, respectively, followed first by a reflection in the $\mathbf{e}_{2'}$-axis and then a rotation of the plane about the origin through the angle $\alpha = \text{arc cos } \frac{24}{25} \approx 16°$.

Example 2. Let **A** be the linear transformation of the space L_3 with matrix

$$\begin{pmatrix} \frac{16}{9} & \frac{2}{9} & \frac{1}{9} \\ \frac{14}{9} & -\frac{14}{9} & \frac{2}{9} \\ -\frac{5}{9} & \frac{14}{9} & \frac{16}{9} \end{pmatrix}$$

in some orthonormal basis $\mathbf{e}_1, \mathbf{e}_2, \mathbf{e}_3$. Express **A** as a product of an orthogonal transformation and a symmetric transformation.

Solution. The symmetric transformation **A*A** has the matrix

$$A^*A = \begin{pmatrix} \frac{16}{9} & \frac{14}{9} & -\frac{5}{9} \\ \frac{2}{9} & -\frac{14}{9} & \frac{14}{9} \\ \frac{1}{9} & \frac{2}{9} & \frac{16}{9} \end{pmatrix} \begin{pmatrix} \frac{16}{9} & \frac{2}{9} & \frac{1}{9} \\ \frac{14}{9} & -\frac{14}{9} & \frac{2}{9} \\ -\frac{5}{9} & \frac{14}{9} & \frac{16}{9} \end{pmatrix} = \begin{pmatrix} \frac{53}{9} & -\frac{26}{9} & -\frac{4}{9} \\ -\frac{26}{9} & \frac{44}{9} & \frac{22}{9} \\ -\frac{4}{9} & \frac{22}{9} & \frac{29}{9} \end{pmatrix}$$

in the basis $\mathbf{e}_1, \mathbf{e}_2, \mathbf{e}_3$. After a bit of calculation based on equation (5), p. 110, we find that the characteristic equation of **A*A** is

$$\lambda^3 - 14\lambda^2 + 49\lambda - 36 = 0,$$

with roots $\lambda_1 = 1, \lambda_2 = 4, \lambda_3 = 9$. The corresponding orthonormal eigenvectors are

$$\mathbf{e}_{1'} = \left(\frac{1}{3}, \frac{2}{3}, -\frac{2}{3}\right), \quad \mathbf{e}_{2'} = \left(\frac{2}{3}, \frac{1}{3}, \frac{2}{3}\right), \quad \mathbf{e}_{3'} = \left(\frac{2}{3}, -\frac{2}{3}, -\frac{1}{3}\right),$$

and the matrix of **A*A** is just

$$H'^2 = \begin{pmatrix} 1 & 0 & 0 \\ 0 & 4 & 0 \\ 0 & 0 & 9 \end{pmatrix}$$

in the basis $\mathbf{e}_{1'}, \mathbf{e}_{2'}, \mathbf{e}_{3'}$. Thus the required symmetric transformation **H** figuring in the representation (3) has the matrix

$$H' = \begin{pmatrix} 1 & 0 & 0 \\ 0 & 2 & 0 \\ 0 & 0 & 3 \end{pmatrix}$$

in the basis $\mathbf{e}_{1'}, \mathbf{e}_{2'}, \mathbf{e}_{3'}$ and the matrix

$$H = \Gamma^{-1}H'\Gamma = \begin{pmatrix} \frac{1}{3} & \frac{2}{3} & \frac{2}{3} \\ \frac{2}{3} & \frac{1}{3} & -\frac{2}{3} \\ -\frac{2}{3} & \frac{2}{3} & -\frac{1}{3} \end{pmatrix}\begin{pmatrix} 1 & 0 & 0 \\ 0 & 2 & 0 \\ 0 & 0 & 3 \end{pmatrix}\begin{pmatrix} \frac{1}{3} & \frac{2}{3} & -\frac{2}{3} \\ \frac{2}{3} & \frac{1}{3} & \frac{2}{3} \\ \frac{2}{3} & -\frac{2}{3} & -\frac{1}{3} \end{pmatrix}$$

$$= \begin{pmatrix} \frac{7}{3} & -\frac{2}{3} & 0 \\ -\frac{2}{3} & 2 & \frac{2}{3} \\ 0 & \frac{2}{3} & \frac{5}{3} \end{pmatrix}$$

in the original basis $\mathbf{e}_1, \mathbf{e}_2, \mathbf{e}_3$. Noting that

$$A^* = \begin{pmatrix} \frac{16}{9} & \frac{14}{9} & -\frac{5}{9} \\ \frac{2}{9} & -\frac{14}{9} & \frac{14}{9} \\ \frac{1}{9} & \frac{2}{9} & \frac{16}{9} \end{pmatrix}, \qquad (A^*)^{-1} = \begin{pmatrix} \frac{14}{27} & \frac{13}{27} & -\frac{7}{27} \\ \frac{1}{27} & -\frac{29}{54} & \frac{13}{27} \\ -\frac{1}{27} & \frac{1}{27} & \frac{14}{27} \end{pmatrix},$$

we can now find the matrix of the orthogonal transformation $\mathbf{S} = (\mathbf{A}^*)^{-1}\mathbf{H}$, the second of the factors figuring in the representation (3):

$$T = (A^*)^{-1}H = \begin{pmatrix} \frac{14}{27} & \frac{13}{27} & -\frac{7}{27} \\ \frac{1}{27} & -\frac{29}{54} & \frac{13}{27} \\ -\frac{1}{27} & \frac{1}{27} & \frac{14}{27} \end{pmatrix}\begin{pmatrix} \frac{7}{3} & -\frac{2}{3} & 0 \\ -\frac{2}{3} & 2 & \frac{2}{3} \\ 0 & \frac{2}{3} & \frac{5}{3} \end{pmatrix}$$

$$= \begin{pmatrix} \frac{8}{9} & \frac{4}{9} & -\frac{1}{9} \\ \frac{4}{9} & -\frac{7}{9} & \frac{4}{9} \\ -\frac{1}{9} & \frac{4}{9} & \frac{8}{9} \end{pmatrix}.$$

Thus the transformation **A** consists of three simultaneous expansions along the $\mathbf{e}_{1'}$, $\mathbf{e}_{2'}$ and $\mathbf{e}_{3'}$-axes, with coefficients 1, 2 and 3, respectively, followed by the (improper) orthogonal transformation with matrix T.

PROBLEMS

1. Use a slight modification of the proof of the representation (3) to prove the alternative representation

$$\mathbf{A} = \mathbf{HT}, \tag{3'}$$

of a nonsingular linear transformation **A**, where **T** is again orthogonal and **H** symmetric.

2. Represent the transformation **A** with each of the following matrices (in some orthonormal basis) in the form (3):

a) $A = \begin{pmatrix} \sqrt{3}+1 & -1 \\ 1 & \sqrt{3}-1 \end{pmatrix}$; b) $A = \begin{pmatrix} 1 & -4 \\ 1 & 4 \end{pmatrix}$.

3. Represent the transformation **A** with matrix

$$A = \begin{pmatrix} 4 & -2 & 2 \\ 4 & 4 & -1 \\ -2 & 4 & 2 \end{pmatrix}$$

(in some orthonormal basis) in the form (3′).

SELECTED HINTS AND ANSWERS

Chapter 1

Sec. 1

1. a) and c) are not linear spaces; b) is a linear space if the line goes through the origin of coordinates.

2. a), c), and d) are linear spaces; b), e), and f) are not.

3. No.

4. Yes. The zero element in R^+ is the number $1 \in R^+$, while the negative of an element $p \in R^+$ is the element $1/p \in R^+$.

7. The set of vectors of L_3 lying in any plane or line going through the origin of coordinates, the space L_3 itself, and the space $\{\mathbf{0}\}$ consisting of the single element $\mathbf{0}$.

8. The sets a), c), and d).

Sec. 2

1. a) $\alpha = -2$; b) $\alpha = -1$; c) $\alpha = \pm 1$; d) $\alpha = 3$, $\beta = 2$; e) $\alpha = -\frac{9}{5}$, $\beta = -\frac{23}{5}$.

2. a) $\alpha = -2$; b) $\alpha = \frac{7}{5}$.

4. Consider the relation $c_1\varphi_1(t) + c_2\varphi_2(t) = 0$ for $t = \frac{1}{2}$ and $t = \frac{3}{2}$.

5. Use the fact that an equation of degree n can have no more than n roots.

6. The functions $1, t, t^2, \ldots, t^n \in C[a, b]$ are linearly independent for arbitrary n (see Prob. 5).

7. Write the equation $\alpha_1\mathbf{a}_1 + \alpha_2\mathbf{a}_2 + \alpha_3\mathbf{a}_3$ in component form, and show that the resulting system of homogeneous equations has a nonzero solution.

9. Since $\mathbf{a}_1, \mathbf{a}_2, \mathbf{a}_3$ are linearly independent, $\alpha(\mathbf{a}_1 + \mathbf{a}_2) + \beta(\mathbf{a}_2 + \mathbf{a}_3) + \gamma(\mathbf{a}_3 + \mathbf{a}_1) = \mathbf{0}$ implies $\alpha + \gamma = \alpha + \beta = \beta + \gamma = 0$ and hence $\alpha = \beta = \gamma = 0$.

Sec. 3

1. $\mathbf{x} = \mathbf{a}_1 + 2\mathbf{a}_2 + 3\mathbf{a}_3$.
Hint. Having proved the linear independence of the vectors $\mathbf{a}_1, \mathbf{a}_2$ and $\mathbf{a}_3$ (see Sec. 2, Prob. 7), write $\mathbf{x}$ in the form $\mathbf{x} = \alpha_1\mathbf{a}_1 + \alpha_2\mathbf{a}_2 + \alpha_3\mathbf{a}_3$. Then write this relation in component form and solve the resulting system for α_1, α_2 and α_3.

2. The dimension equals $n + 1$, the simplest basis consisting of the polynomials $1, t, t^2, \ldots, t^n$. The components of a polynomial $P(t) = a_0 + a_1t + a_2t^2 + \ldots + a_nt^n$ in this basis are just the coefficients $a_0, a_1, a_2, \ldots, a_n$.

3. Infinite-dimensional because of the result of Sec. 2, Prob. 6.

4. One-dimensional with any element $x \neq 1$ as a basis.

5. The result follows from the fact that the basis for L' is also a basis for L.

6. Choose a basis in $L' \cap L''$, and enlarge it to make first a basis for L' and then a basis for L''. Then prove that the vectors of the basis in $L' \cap L''$ together with both sets of supplementary vectors form a basis in $L' + L''$.

7. Use the results of Probs. 5 and 6.

8. Use the result of Prob. 6.

9. The sum is the whole space L_3, while the intersection is one-dimensional (a straight line).

10. $s = 3$, $d = 2$.

11. A basis for $L' + L''$ is given, say, by the vectors $\mathbf{a}_1, \mathbf{a}_2, \mathbf{a}_3, \mathbf{b}_2$ and a basis for $L' \cap L''$ by the vectors $\mathbf{b}_1 = -2\mathbf{a}_1 + \mathbf{a}_2 + \mathbf{a}_3$, $\mathbf{b}_3 = 5\mathbf{a}_1 - \mathbf{a}_2 - 2\mathbf{a}_3$.

12. a) A basis is given, say, by the vectors $(1, 1, 0, \ldots, 0)$, $(0, 0, 1, 0, \ldots, 0)$, $(0, 0, 0, 1, 0, \ldots, 0), \ldots, (0, 0, 0, \ldots, 1)$. The dimension equals $n - 1$; b) A basis is given, say, by the vectors $(1, 0, \ldots, 0)$, $(0, 0, 1, 0, \ldots, 0)$, $(0, 0, 0, 0, 1, 0, \ldots, 0)$, and the vector $(0, 1, 0, 1, 0, 1, \ldots)$. The dimension equals $1 + [\frac{1}{2}(n + 1)]$, where $[\frac{1}{2}(n + 1)]$ denotes the largest integer not exceeding $\frac{1}{2}(n + 1)$. c) A basis is given, say, by the vectors $(1, 0, 1, 0, \ldots)$ and $(0, 1, 0, 1, \ldots)$. The dimension equals 2. d) A basis is given, say, by the vectors $(1, 0, 0, \ldots, -1), (0, 1, 0, \ldots, -1), \ldots, (0, 0, \ldots, 1, -1)$. The dimension equals $n - 1$.

13. Any n linearly independent solutions of the equation form a basis, and the space is of dimension n. The components of an arbitrary solution with respect to any basis are just the coefficients of the expansion of the solution with respect to the elements of the basis.

14. $\mathbf{a}_1x_1 + \mathbf{a}_2x_2 + \cdots + \mathbf{a}_nx_n = b$, where $\mathbf{a}_i = (a_{1i}, \ldots, a_{mi})$ $(i = 1, \ldots, n)$ and $\mathbf{b} = (b_1, \ldots, b_m)$ are vectors of the space L_m.

Sec. 4

1. In the triangle ABC write $\overrightarrow{BC}$ in the form $\overrightarrow{AC} - \overrightarrow{AB}$, and then find $|\overrightarrow{BC}|^2$. b) In the parallelogram $ABCD$ we have $\overrightarrow{AC} = \overrightarrow{AB} + \overrightarrow{BC}$, $\overrightarrow{BD} = \overrightarrow{BC} - \overrightarrow{AB}$. Now find $|\overrightarrow{AC}|^2 + |\overrightarrow{BD}|^2$. c) In the rhombus $ABCD$ we have $|\overrightarrow{AB}|^2 = |\overrightarrow{AD}|^2$, and hence $\overrightarrow{AC} \cdot \overrightarrow{DB} = (\overrightarrow{AB} - \overrightarrow{AD}) \cdot (\overrightarrow{AB} + \overrightarrow{AD}) = 0$. d) In the rectangle $ABCD$ we have $\overrightarrow{AB} \cdot \overrightarrow{BC} = 0$, and hence $|\overrightarrow{AB} + \overrightarrow{BC}|^2 = |\overrightarrow{AB} - \overrightarrow{BC}|^2$, i.e., $|\overrightarrow{AC}|^2 = |\overrightarrow{BD}|^2$ or $|\overrightarrow{AC}| = |\overrightarrow{BD}|$. e) The proof is similar to a). f) The median AD of the triangle ABC is given by $\overrightarrow{AD} = \frac{1}{2}(\overrightarrow{AC} + \overrightarrow{AB})$. Now find $|\overrightarrow{AD}|^2$, using the result of a). g) Let AA_1 and BB_1 be equal medians of the triangle ABC. Then $|\overrightarrow{AA_1}|^2 = |\overrightarrow{BB_1}|^2$, and hence $|\overrightarrow{AB} + \overrightarrow{AC}|^2 = |\overrightarrow{BA} + \overrightarrow{BC}|^2$ or $(\overrightarrow{AB} + \overrightarrow{AC} + \overrightarrow{BA} + \overrightarrow{BC}) \cdot (\overrightarrow{AB} + \overrightarrow{AC} - \overrightarrow{BA} - \overrightarrow{BC}) = 0$, so that $\overrightarrow{AB} \cdot \overrightarrow{CC_1} = 0$ where CC_1 is the other median of the triangle. h) In the trapezoid $ABCD$ we have $\overrightarrow{AB} + \overrightarrow{BC} + \overrightarrow{CD} = \overrightarrow{AD}$, $\overrightarrow{AC} = \overrightarrow{AB} + \overrightarrow{BC}$, $\overrightarrow{BD} = \overrightarrow{BA} + \overrightarrow{AD}$, and hence $|\overrightarrow{AC}|^2 + |\overrightarrow{BD}|^2 = |\overrightarrow{AB} + \overrightarrow{BC}|^2 + |\overrightarrow{BA} + \overrightarrow{AD}|^2 = |\overrightarrow{AB}|^2 + 2\overrightarrow{AB} \cdot \overrightarrow{BC} + |\overrightarrow{BC}|^2 + |\overrightarrow{AB}|^2 - 2\overrightarrow{AB} \cdot \overrightarrow{AD} + |\overrightarrow{AD}|^2 = |\overrightarrow{AD}|^2 + |\overrightarrow{BC}|^2 + 2\{|\overrightarrow{AB}|^2 + \overrightarrow{AB} \cdot (\overrightarrow{BC} - \overrightarrow{AD})\} = |\overrightarrow{AD}|^2 + |\overrightarrow{BC}|^2 + 2\overrightarrow{AB} \cdot (\overrightarrow{AB} + \overrightarrow{BC} - \overrightarrow{AD}) = |\overrightarrow{AD}|^2 + |\overrightarrow{BC}|^2 + 2\overrightarrow{AB} \cdot \overrightarrow{DC} = |\overrightarrow{AD}|^2 + |\overrightarrow{BC}|^2 + 2|\overrightarrow{AB}||\overrightarrow{DC}|$. i) In a regular tetrahedron $A_1A_2A_3A_4$ we have $\overrightarrow{A_3A_4} = \overrightarrow{A_1A_4} - \overrightarrow{A_1A_3}$, $\overrightarrow{A_1A_2} \cdot \overrightarrow{A_3A_4} = \overrightarrow{A_1A_2} \cdot \overrightarrow{A_1A_4} - \overrightarrow{A_1A_2} \cdot \overrightarrow{A_1A_3}$ or $\overrightarrow{A_1A_2} \cdot \overrightarrow{A_3A_4} = l^2 \cos 60° - l^2 \cos 60° = 0$, where l is the length of a side of the tetrahedron, i.e., $\overrightarrow{A_1A_2} \cdot \overrightarrow{A_3A_4} = 0$.

2. $(x_i y_i)^2 \leq (x_j x_j)(y_k y_k)$.

4. a) Yes; b) Yes; c) No.

6. $\sqrt{(f(t), f(t))} = \sqrt{\int_a^b f^2(t)\,dt}$.

8. Start from the arbitrary inequality $(\lambda \mathbf{x} - \mathbf{y}) \cdot (\lambda \mathbf{x} - \mathbf{y}) \geq 0$.

9. In E_n the inequality is the same as in Prob. 2, except that now $i, j, k = 1, 2, \ldots, n$ instead of $i, j, k = 1, 2, 3$. In $C[a, b]$ we have

$$\left| \int_a^b f(t)g(t)\,dt \right| \leq \sqrt{\int_a^b f^2(t)\,dt} \sqrt{\int_a^b g^2(t)\,dt}.$$

10. 30°, 90°, 120°.

14. Take the scalar product of the vector $\mathbf{x}_1 + \mathbf{x}_2 + \cdots + \mathbf{x}_k$ with itself.

15. $$|\mathbf{x} + \mathbf{y}|^2 = \mathbf{x} \cdot \mathbf{x} + 2\mathbf{x} \cdot \mathbf{y} + \mathbf{y} \cdot \mathbf{y} \begin{cases} \leq |\mathbf{x}|^2 + 2|\mathbf{x}||\mathbf{y}| + |\mathbf{y}|^2, \\ \geq |\mathbf{x}|^2 - 2|\mathbf{x}||\mathbf{y}| + |\mathbf{y}|^2, \end{cases}$$
by the Cauchy—Schwarz inequality.

16. $$\left| \sqrt{\int_a^b f^2(t)\,dt} - \sqrt{\int_a^b g^2(t)\,dt} \right| \leq \sqrt{\int_a^b [f(t) + g(t)]^2\,dt} \leq \sqrt{\int_a^b f^2(t)\,dt} + \sqrt{\int_a^b g^2(t)\,dt}.$$

17. Calculate

$$\left| \mathbf{x} - \sum_{i=1}^{k} (\mathbf{x}\cdot\mathbf{e}_i)\mathbf{e}_i \right|^2 \geq 0,$$

noting that

$$\mathrm{Pr}_{\mathbf{e}_i}\mathbf{x} = \mathbf{x}\cdot\mathbf{e}_i, \qquad \mathbf{x} = \sum_{i=1}^{n} (\mathbf{x}\cdot\mathbf{e}_i)\mathbf{e}_i.$$

18. a) Let $u_k(t) = (t^2 - 1)^k$, and prove that $u_k^{(j)}(1) = 0$ if $j < k$. Then integrate $\int_{-1}^{1} u_k^{(k)}(t)t^j\,dt$ by parts repeatedly until the integrand no longer contains a power of t. Show that this integral vanishes if $j = 0, 1, \ldots, k - 1$, thereby deducing the orthogonality of the Legendre polynomials. b) $P_0(t) = 1, P_1(t) = t$, $P_2(t) = \frac{1}{2}(3t^2 - 1)$, $P_3(t) = \frac{1}{2}(5t^3 - 3t)$, $P_4(t) = \frac{1}{8}(35t^4 - 30t^2 + 3)$, $P_k(t) = \dfrac{1}{2^k k!}\sum_{j=0}^{k} (-1)^{k-j}\binom{k}{j}\dfrac{(2j)!}{(2j-k)!}t^{2j-k} = \sum_{j=0}^{k} (-1)^{k-j}\dfrac{1\cdot3\cdot5\cdots(2j-1)}{(k-j)!(2j-k)!2^{k-j}}t^{2j-k}$, where the terms with negative powers of t must be dropped. c) $\sqrt{\dfrac{2}{2k+1}}$.

Hint. Writing $(t^2 - 1)^k = u_k(t)$, show that $\int_{-1}^{1} u_k^{(k)}(t)u_k^{(k)}(t)\,dt = (k!)^2\dfrac{2^{k+1}}{2k+1}$.

d) $P_k(1) = 1$.
Hint. Use Leibniz's rule for differentiating a product.

Sec. 5

1. $|\mathbf{a} \times (\mathbf{b} + \mathbf{c})|$, $|\mathbf{b} \times (\mathbf{a} + \mathbf{c})|$, $|\mathbf{c} \times (\mathbf{a} + \mathbf{b})|$.

2. $\sin\alpha = \dfrac{|\overrightarrow{OA} \times \overrightarrow{OB} + \overrightarrow{OB} \times \overrightarrow{OC} + \overrightarrow{OC} \times \overrightarrow{OA}|}{|\overrightarrow{OB} - \overrightarrow{OA}||\overrightarrow{OC} - \overrightarrow{OA}|}$.

3. $h_1 = \dfrac{|(\mathbf{r}_1 - \mathbf{r}_2) \times (\mathbf{r}_3 - \mathbf{r}_2)|}{|\mathbf{r}_3 - \mathbf{r}_2|}$, etc.

4. We have $\mathbf{n}_1 = \mathbf{r}_2 \times \mathbf{r}_1$, $\mathbf{n}_2 = \mathbf{r}_3 \times \mathbf{r}_2$, $\mathbf{n}_3 = \mathbf{r}_1 \times \mathbf{r}_3$, $\mathbf{n}_4 = (\mathbf{r}_2 - \mathbf{r}_1) \times (\mathbf{r}_3 - \mathbf{r}_1)$, where $\mathbf{r}_1 = \overrightarrow{OA}$, $\mathbf{r}_2 = \overrightarrow{OB}$, $\mathbf{r}_3 = \overrightarrow{OC}$, and hence $\mathbf{n}_1 + \mathbf{n}_2 + \mathbf{n}_3 + \mathbf{n}_4 = \mathbf{0}$. It follows that $|\mathbf{n}_4|^2 = |\mathbf{n}_1 + \mathbf{n}_2 + \mathbf{n}_3|^2$, which implies $S_4^2 = S_1^2 + S_2^2 + S_3^2 + 2\mathbf{n}_1\cdot\mathbf{n}_2 + 2\mathbf{n}_2\cdot\mathbf{n}_3 + 2\mathbf{n}_3\cdot\mathbf{n}_1$. Now use the fact that the cosine of the angle between two faces differs only in sign from the cosine of the angle between the normals to the faces.

7. $\begin{vmatrix} a_1 & a_2 \\ b_1 & b_2 \end{vmatrix}^2 + \begin{vmatrix} a_1 & a_3 \\ b_1 & b_3 \end{vmatrix}^2 + \begin{vmatrix} a_2 & a_3 \\ b_2 & b_3 \end{vmatrix}^2 = \begin{vmatrix} a_ia_i & a_ib_i \\ a_ib_i & b_ib_i \end{vmatrix}$, where a_i and b_i are the components of $\mathbf{a}$ and $\mathbf{b}$ in some orthonormal basis.

10. The indicated lines are collinear with the vectors $\mathbf{r}_1 \times (\mathbf{r}_2 \times \mathbf{r}_3)$, $\mathbf{r}_2 \times (\mathbf{r}_3 \times \mathbf{r}_1)$ and $\mathbf{r}_3 \times (\mathbf{r}_1 \times \mathbf{r}_2)$, where $\mathbf{r}_1$, $\mathbf{r}_2$, and $\mathbf{r}_3$ are vectors collinear with the edges of the angle. Now use the result of Prob. 9.

11. Use the result of Prob. 6 repeatedly to prove that $\mathbf{p}\cdot\mathbf{q} = 0$.

12. $S = \frac{1}{2}|(\mathbf{b} - \mathbf{a}) \times (\mathbf{c} - \mathbf{a})|$, and hence $4S^2 = b^2c^2 \sin^2\alpha + a^2c^2 \sin^2\beta + a^2b^2 \sin^2\gamma + 2abc^2 (\cos\alpha \cos\beta - \cos\gamma) + 2bca^2 (\cos\beta \cos\gamma - \cos\alpha) + 2acb^2 (\cos\alpha \cos\gamma - \cos\beta)$.

13. 2(**a**, **b**, **c**). The volume of the parallelepiped constructed on the diagonals of three faces passing through one vertex, equal to twice the volume of the original parallelepiped.

14. $\lambda\mu\nu = -1$.

15. Write the system in vector form (cf. Sec. 3, Prob. 14), and then take the scalar product of both sides with $\mathbf{a}_2 \times \mathbf{a}_3$, $\mathbf{a}_3 \times \mathbf{a}_1$ and $\mathbf{a}_1 \times \mathbf{a}_2$.

17. a) Calculate $(\mathbf{a} \times \mathbf{b}) \times (\mathbf{c} \times \mathbf{d})$ in two different ways and compare the results. b) Use the result of Prob. 16a. The formula means that the volume of the parallelepiped whose edges are perpendicular to the faces of the original parallelepiped and have numerical values equal to the areas of these faces is equal to the square of the volume of the original parallelepiped.

18. By Prob. 17a we have $(\mathbf{a}, \mathbf{b}, \mathbf{c})\mathbf{d} = (\mathbf{b}, \mathbf{c}, \mathbf{d})\mathbf{a} + (\mathbf{c}, \mathbf{a}, \mathbf{d})\mathbf{b} + (\mathbf{a}, \mathbf{b}, \mathbf{d})\mathbf{c} = [(\mathbf{b} \times \mathbf{c})\cdot\mathbf{d}]\mathbf{a} + [(\mathbf{c} \times \mathbf{a})\cdot\mathbf{d}]\mathbf{b} + [(\mathbf{a} \times \mathbf{b})\cdot\mathbf{d}]\mathbf{c}$. Now replace **d** by $\mathbf{x} \times \mathbf{y}$, use Prob. 6, and take the scalar product of both sides of the resulting equation with **z**.

19. Use Prob. 18.

Sec. 6

1. a) $\mathbf{e}_{1'} = -\mathbf{e}_1 \sin\theta + \mathbf{e}_2 \cos\theta$, $\mathbf{e}_{2'} = \mathbf{e}_1 \cos\theta + \mathbf{e}_2 \sin\theta$, $x_{1'} = -x_1 \sin\theta + x_2 \cos\theta$, $x_{2'} = x_1 \cos\theta + x_2 \sin\theta$; b) $x_{1'} = -x_1$, $x_{2'} = x_2$.

2. a) $\Gamma = \begin{pmatrix} 0 & 1 & 0 \\ 1 & 0 & 0 \\ 0 & 0 & 1 \end{pmatrix}$; b) $\Gamma = \begin{pmatrix} 0 & 0 & 1 \\ 1 & 0 & 0 \\ 0 & 1 & 0 \end{pmatrix}$.

3. a) Two rows are interchanged; b) Two columns are interchanged; c) The new matrix is obtained by reflecting the old matrix in its central term.

4. $\mathbf{e}_{1'} = \mathbf{e}_1(\cos\varphi\cos\psi - \sin\varphi\sin\psi\cos\theta) + \mathbf{e}_2(\sin\varphi\cos\psi + \cos\varphi\sin\psi\cos\theta) + \mathbf{e}_3 \sin\psi\sin\theta$, $\mathbf{e}_{2'} = \mathbf{e}_1(-\cos\varphi\sin\psi - \sin\varphi\cos\psi\cos\theta) + \mathbf{e}_2(\cos\theta\cos\varphi\cos\psi - \sin\varphi\sin\psi) + \mathbf{e}_3\cos\psi\sin\theta$, $\mathbf{e}_{3'} = \mathbf{e}_1 \sin\varphi\sin\theta - \mathbf{e}_2\cos\varphi\sin\theta + \mathbf{e}_3\cos\theta$.

6. $\mathbf{e}_{n'} = \dfrac{\mathbf{x}}{|\mathbf{x}|}$, while $\mathbf{e}_{1'}, \mathbf{e}_{2'}, \ldots, \mathbf{e}_{(n-1)'}$ are arbitrary.

7. Choose a new basis $\mathbf{e}_{1'}, \ldots, \mathbf{e}_{n'}$ whose first k vectors form a basis for L_k. Write the condition that a vector **x** belong to L_k as a system of equations in the new basis, and then write the corresponding system in the old basis.

8. $\begin{pmatrix} 1 & 0 & \cdots & 0 \\ -a & 1 & \cdots & 0 \\ a^2 & -2a & \cdots & 0 \\ \cdot & \cdot & \cdots & \cdot \\ (-1)^n a^n & (-1)^{n-1} n a^{n-1} & \cdots & 1 \end{pmatrix}$,

where the $(k+1)$st row of the matrix consists of the numbers $(-a)^k$, $C_{k-1}^k(-a)^{k-1}$, $C_{k-2}^k(-a)^{k-2}, \ldots, C_1^k(-a), 1, \underbrace{0, \ldots, 0}_{n-k \text{ times}}$.†

---

†C_k^n denotes the binomial coefficient $n!/k!(n-k)!$, where $C_k^n = 0$ if $k \leq 0$.

Sec. 7

1. a) $(\mathbf{x}-\mathbf{x}_1, \mathbf{a}, \mathbf{b}) = 0$, $\begin{vmatrix} x_1 - x_1^{(1)} & x_2 - x_2^{(1)} & x_3 - x_3^{(1)} \\ a_1 & a_2 & a_3 \\ b_1 & b_2 & b_3 \end{vmatrix} = 0$, where $\mathbf{x} = x_i\mathbf{e}_i$, $\mathbf{x}_1 = x_i^{(1)}\mathbf{e}_i$, $\mathbf{a} = a_i\mathbf{e}_i$, $\mathbf{b} = b_i\mathbf{e}_i$; b) $(\mathbf{x}-\mathbf{x}_0, \mathbf{x}_1-\mathbf{x}_0, \mathbf{a}) = 0$,

$$\begin{vmatrix} x_1 - x_1^{(0)} & x_2 - x_2^{(0)} & x_3 - x_3^{(0)} \\ x_1^{(1)} - x_1^{(0)} & x_2^{(1)} - x_2^{(0)} & x_3^{(1)} - x_3^{(0)} \\ a_1 & a_2 & a_3 \end{vmatrix} = 0,$$ where $\mathbf{x} = x_i\mathbf{e}_i$, $\mathbf{x}_0 = x_i^{(0)}\mathbf{e}_i$, $\mathbf{x}_1 = \mathbf{x}_i^{(1)}\mathbf{e}_i$, $\mathbf{a} = a_i\mathbf{e}_i$.

2. The planes intersect if (and only if) some

$$A_{ij} = \begin{vmatrix} a_i^{(1)} & a_i^{(2)} \\ a_j^{(1)} & a_j^{(2)} \end{vmatrix} \neq 0,$$

are parallel if all

$$A_{ij} = 0, \qquad \frac{a_i^{(1)}}{b^{(1)}} \neq \frac{a_i^{(2)}}{b^{(2)}},$$

and are coincident if all

$$A_{ij} = 0, \qquad \frac{a_i^{(1)}}{b^{(1)}} = \frac{a_i^{(2)}}{b^{(2)}}.$$

3. $\dfrac{|b - b'|}{\sqrt{a_i a_i}}$.

4. $a_i x_i + \frac{1}{2}(b + b') = 0$.

5. $\lambda(a_i^{(1)}x_i + b^{(1)}) + \mu(a_i^{(2)}x_i + b^{(2)}) = 0$.

6. a) $(a_i^{(2)}x_i^{(0)} + b^{(2)})(a_i^{(1)}x_i + b^{(1)}) - (a_i^{(1)}x_i^{(0)} + b^{(1)})(a_i^{(2)}x_i + b^{(2)}) = 0$; b) $a_k^{(2)}a_k^{(3)}(a_i^{(1)}x_i + b^{(1)} - a_k^{(1)}a_k^{(3)})(a_i^{(2)}x_i + b^{(2)}) = 0$.

7. $\cos\theta = \dfrac{a_i^{(1)}a_i^{(2)}}{\sqrt{a_i^{(1)}a_i^{(1)}}\sqrt{a_i^{(2)}a_i^{(2)}}}$; the planes are orthogonal when $a_i^{(1)}a_i^{(2)} = 0$.

8. Choose $\lambda = \pm a^{(1)}(a^{(2)})^2 - a^{(2)}a_i^{(1)}a_i^{(2)}$, $\mu = a^{(2)}(a^{(1)})^2 \mp a^{(1)}a_i^{(1)}a_i^{(2)}$ in the answer to Prob. 5, where $a^{(1)} = \sqrt{a_i^{(1)}a_i^{(1)}}$, $a^{(2)} = \sqrt{a_i^{(2)}a_i^{(2)}}$.

9. $x_i^{(0)} - (b + a_j x_j^{(0)})\dfrac{a_i}{\sqrt{a_k a_k}}$.

10. $\frac{1}{2}\sqrt{\epsilon_{ijp}\epsilon_{klp}(z_i - x_i)(z_j - x_j)(y_k - x_k)(y_l - x_l)}$.

11. $\frac{1}{6}\epsilon_{ijk}(u_i - x_i)(z_j - x_j)(y_k - x_k)$.

12. $\dfrac{|\mathbf{a} \times (\mathbf{x}_0 - \mathbf{y})|}{|\mathbf{a}|}$.

13. $\dfrac{|\mathbf{a} \times (\mathbf{x}_2 - \mathbf{x}_1)|}{|\mathbf{a}|}$.

14. a) $\arccos \dfrac{\mathbf{a}_1 \cdot \mathbf{a}_2}{|\mathbf{a}_1||\mathbf{a}_2|}$; b) $\dfrac{|(\mathbf{x}_2 - \mathbf{x}_1, \mathbf{a}_1, \mathbf{a}_2)|}{|\mathbf{a}_1 \times \mathbf{a}_2|}$.

Chapter 2

Sec. 8

1. a), d), and e) are linear forms, but not b); c) is a linear form only if $c = 0$.

2. $\varphi(\mathbf{x}) = (\mathbf{a}, \mathbf{b}, \mathbf{x}) = (\mathbf{a} \times \mathbf{b}) \cdot \mathbf{x}$.

Sec. 9

3. Yes.

4. Yes.

5. No.

6. No, unless $c = 0$.

Sec. 10

1. No, unless $c = 0$.

2. No.

3. Yes.

4. Yes.

7, 8. Examine the cases where the components equal 0, 1 separately.

10. $\dfrac{\partial \varphi}{\partial x_{1'}} = \dfrac{\partial \varphi}{\partial x_i} \dfrac{\partial x_i}{\partial x_{i'}} = \gamma_{i'i} \dfrac{\partial \varphi}{\partial x_i}$,
since $x_i = \gamma_{ii'} x_{i'}$ implies

$$\frac{\partial x_i}{\partial x_{i'}} = \gamma_{ii'} = \gamma_{i'i}.$$

Similarly,

$$\frac{\partial^2 \varphi}{\partial x_{i'} \partial x_{j'}} = \frac{\partial}{\partial x_i}\left(\frac{\partial \varphi}{\partial x_{j'}}\right) \frac{\partial x_i}{\partial x_{i'}} = \frac{\partial}{\partial x_i}\left(\frac{\partial \varphi}{\partial x_j} \gamma_{jj'}\right) \gamma_{ii'} = \gamma_{i'i} \gamma_{j'j} \frac{\partial^2 \varphi}{\partial x_i \partial x_j}.$$

Sec. 11

2. $a_{ijk} b_{lm}$; $a_{ijk} b_{im}$, $a_{ijk} b_{li}$, $a_{ijk} b_{jm}$, $a_{ijk} b_{lj}$, $a_{ijk} b_{km}$, $a_{ijk} b_{lk}$; $a_{ijk} b_{ij}$, $a_{ijk} b_{ji}$, $a_{ijk} b_{jk}$, $a_{ijk} b_{kj}$, $a_{ijk} b_{ik}$, $a_{ijk} b_{ki}$.

3. To prove the sufficiency, write the condition in the form

$$\frac{z_{ij}}{z_{il}} = \frac{z_{kj}}{z_{kl}} = \lambda_{jl},$$

which implies $z_{ij} = z_{il}$ (no summation over l). Now let $l = 1$.

4. $a_{ii} = 1$.

5. a) (16, 19, 41); b) (25, 21, 36); c) (37, 2, 16); d) (3, 20, 40); e) 186; f) 140; g) 10; h) $\begin{pmatrix} -2 & 0 & 3 \\ 5 & -3 & 2 \\ 4 & 5 & 3 \end{pmatrix}$; i) (17, 17, 20); j) 150.

6. For example, such a basis consists of the nine tensors with matrices of the form

$$\begin{pmatrix} 1 & 0 & 0 \\ 0 & 0 & 0 \\ 0 & 0 & 0 \end{pmatrix}, \begin{pmatrix} 0 & 1 & 0 \\ 0 & 0 & 0 \\ 0 & 0 & 0 \end{pmatrix}, \ldots, \begin{pmatrix} 0 & 0 & 0 \\ 0 & 0 & 0 \\ 0 & 0 & 1 \end{pmatrix}.$$

Sec. 12

1. Prove that the coefficients a_{ijk} of the form φ are proportional to the components of the discriminantal tensor ϵ_{ijk}.

2. Use the fact that two of the indices of the tensor determined by such a form are always equal.

4. A consequence of the fact that $a_{ijk} = a_{jik} = -a_{jki} = -a_{kji} = a_{kij} = a_{ikj} = -a_{ijk}$.

5. Consider the terms with $i = j$ and $i \neq j$ separately.

6. Collect similar terms, set the coefficients of distinct $x_i x_j x_k$ equal to zero, and use the symmetry of a_{ijk} in the first two indices.

8. In the first part, collect similar terms and set the coefficients of distinct $x_i y_j x_k y_l$ equal to zero.

12. a) 6; b) 0; c) 0; d) (−1, 2, 1); e) 0; f) 9; g) (4, 41, −3); h) 143.

13. a) $x_1^2 + x_2^2 + x_3^2 = 1/\lambda$; a sphere of radius $\sqrt{1/\lambda}$ (real or imaginary depending on the sign of λ); b) $(a_i x_i)(b_j x_j) = 1$ or $x_{1'} x_{2'} = c$ after an appropriate coordinate transformation; a hyperbolic cylinder.

14. a) $x_1^3 + x_2^3 = 1$; b) $x_1 x_2 (x_1 + x_2) = 1$; c) $x_1^3 - 3x_1 x_2^2 = 1$.

Chapter 3

Sec. 13

2. a), b), c) and d) are linear, but e) is not; a) represents reflection in the origin, b) carries the vector $\mathbf{x}$ into the vector lying on the bisector of the first and third quadrants with the same first component as $\mathbf{x}$; c) represents twofold expansion of $\mathbf{x}$ along $\mathbf{e}_2$ followed by reflection in $\mathbf{e}_1$, d) represents expansion λ_1 times along $\mathbf{e}_1$ followed by expansion λ_2 times along $\mathbf{e}_2$ (if $\lambda_1 < 0$, the first expansion must be accompanied by reflection in $\mathbf{e}_2$, and similarly if $\lambda_2 < 0$).

3. $\mathbf{u} = \mathbf{A}\mathbf{x} = \lambda x_1 \mathbf{e}_1 + x_2 \mathbf{e}_2$.

5. a), d), e), and f) are linear, b) and g) are nonlinear, c) is linear only if $\mathbf{a} = \mathbf{0}$; a) represents projection onto $\mathbf{a}$ followed by expansion a^2 times, d) represents projection onto the $\mathbf{e}_1, \mathbf{e}_2$-plane, e) represents reflection in the $\mathbf{e}_1, \mathbf{e}_3$-plane followed by reflection in the $\mathbf{e}_1, \mathbf{e}_2$-plane and twofold expansion along $\mathbf{e}_3$, f) represents expansion (or contraction) along $\mathbf{e}_3$.

6. Yes.

12. Only the operation c) provided that $H(t, s)$ is a polynomial of degree not exceeding n in t with coefficients which are functions of s.

Sec. 14

1. $A = \begin{pmatrix} -1 & 0 \\ 0 & -1 \end{pmatrix}$ in Prob. 2a; $A = \begin{pmatrix} 1 & 0 \\ 1 & 0 \end{pmatrix}$ in Prob. 2b; $A = \begin{pmatrix} 1 & 0 \\ 0 & -2 \end{pmatrix}$ in Prob. 2c; $A = \begin{pmatrix} \lambda_1 & 0 \\ 0 & \lambda_2 \end{pmatrix}$ in Prob. 2d; $A = \begin{pmatrix} \lambda & 0 \\ 0 & 1 \end{pmatrix}$ in Prob. 3.

3. $A = \begin{pmatrix} a_1^2 & a_1a_2 & a_1a_3 \\ a_2a_1 & a_2^2 & a_2a_3 \\ a_3a_1 & a_3a_2 & a_3^2 \end{pmatrix}$ in Prob. 5a, where $\mathbf{a} = a_i\mathbf{e}_i$; $A = N$ for $\mathbf{a} = \mathbf{0}$ in Prob. 5c; $A = \begin{pmatrix} 1 & 0 & 0 \\ 0 & 1 & 0 \\ 0 & 0 & 0 \end{pmatrix}$ in Prob. 5d; $A = \begin{pmatrix} 1 & 0 & 0 \\ 0 & -1 & 0 \\ 0 & 0 & -2 \end{pmatrix}$ in Prob. 5e; $A = \begin{pmatrix} 1 & 0 & 0 \\ 0 & 1 & 0 \\ 0 & 0 & \lambda \end{pmatrix}$ in Prob. 5f. In Prob. 6, $A = \epsilon \begin{pmatrix} 0 & -a_3 & a_2 \\ a_3 & 0 & -a_1 \\ -a_2 & a_1 & 0 \end{pmatrix}$, where $\mathbf{a} = a_i\mathbf{e}_i$ and $\epsilon = \pm 1$ depending on whether the basis is right-handed or left-handed. In Prob. 7, $A = \begin{pmatrix} \lambda_1 & 0 & 0 \\ 0 & \lambda_2 & 0 \\ 0 & 0 & \lambda_3 \end{pmatrix}$. In Prob. 8, $A = \begin{pmatrix} \frac{1}{3} & \frac{1}{3} & \frac{1}{3} \\ \frac{1}{3} & \frac{1}{3} & \frac{1}{3} \\ \frac{1}{3} & \frac{1}{3} & \frac{1}{3} \end{pmatrix}$. In Prob. 9, $A = \begin{pmatrix} 0 & 0 & 1 \\ 1 & 0 & 0 \\ 0 & 1 & 0 \end{pmatrix}$ if $\mathbf{e}_1$ goes into $\mathbf{e}_2$, while $A = \begin{pmatrix} 0 & 1 & 0 \\ 0 & 0 & 1 \\ 1 & 0 & 0 \end{pmatrix}$ if $\mathbf{e}_2$ goes into $\mathbf{e}_1$.

5. a) $A = \begin{pmatrix} 0 & 1 & 0 \cdots 0 \\ 0 & 0 & 2 \cdots 0 \\ \cdot & \cdot & \cdots\cdot \\ 0 & 0 & 0 \cdots n \\ 0 & 0 & 0 \cdots 0 \end{pmatrix}$; b) $A = \begin{pmatrix} 0 & 1 & 0 \cdots 0 \\ 0 & 0 & 1 \cdots 0 \\ \cdot & \cdot & \cdots\cdot \\ 0 & 0 & 0 \cdots 1 \\ 0 & 0 & 0 \cdots 0 \end{pmatrix}$.

6. If $\mathbf{a}_i = a_{ij}\mathbf{e}_j$, $\mathbf{b}_i = b_{ij}\mathbf{e}_j$ and $A = (a_{ij})$, $B = (b_{ij})$, then the transformation matrix C has elements

$$c_{ij} = \frac{b_{ki}A_{kj}}{|A|},$$

where A_{kj} is the cofactor of a_{jk} and $|A|$ is the determinant of the matrix A.

7. $C = \begin{pmatrix} 2 & -11 & 6 \\ 1 & -7 & 4 \\ 2 & -1 & 0 \end{pmatrix}$.

8. a) Reflection in the $\mathbf{e}_2, \mathbf{e}_3$-plane; b) λ-fold expansion along $\mathbf{e}_2$; c) Projection onto the $\mathbf{e}_1, \mathbf{e}_3$-plane; d) Projection onto $\mathbf{e}_2$.

9. $A = (a_{ik})$ where $a_{ik} = \omega_i\omega_k + (\delta_{ik} - \omega_i\omega_k)\cos\alpha + \epsilon_{ijk}\omega_j \sin\alpha$.

11. The transformations considered in Examples 1, 2, 5, 7, 8 and in Probs. 2a, 2c, 2d, 3, 4, 5e, 5f, 7, and 9 are nonsingular, while those considered in Examples 3, 6 and Probs. 2b, 5a, 5c (if $\mathbf{a} = \mathbf{0}$), 5d, 6, 8, and 10 are singular with matrices of rank 0, 1, 1, 1, 0, 2, 2, 1, n, respectively.

13. a) is a singular transformation with matrix of rank 1 carrying every vector of L_2 into a vector on the line $x_2 = 2x_1$, b) is a singular transformation with matrix of rank 2 carrying every vector of L_3 into a vector in the plane $x_1 + x_2 = x_3$, c) is a singular transformation with matrix of rank 1 carrying every vector of L_3 into a vector on the line $x_1 = \frac{1}{2}x_2 = \frac{1}{3}x_3$.

14. Prove that a)–d) hold if and only if $\mathbf{A}$ has a matrix of rank 3.

17. Use the results of Probs. 14a and 14d.

18. a) The null space consists of the vectors collinear with $\mathbf{e}_2$, the range consists of the vectors collinear with $a\mathbf{e}_1 + \mathbf{e}_2$, the defect and rank both equal 1; b) The null space consists of the vectors collinear with $\mathbf{e}_2$, the range consists of the vectors of the $\mathbf{e}_1, \mathbf{e}_3$-plane, the defect equals 1, the rank equals 2; c) The null space consists of the vectors of the $\mathbf{e}_1, \mathbf{e}_2$-plane, the range consists of the vectors collinear with $\mathbf{e}_3$, the defect equals 2, the rank equals 1; d) The null space consists of the vectors collinear with $\mathbf{e}_2$, the range consists of the $\mathbf{e}_2, \mathbf{e}_3$-plane, the defect equals 1, the rank equals 2.

19. The null space consists of all polynomials of degree 0, the range consists of all polynomials of degree not exceeding $n - 1$, the defect equals 1, the rank equals n.

Sec. 15

1. a) $\varphi = x_1'^2, x_1'^2 = 1$, a pair of lines parallel to $\mathbf{e}_2$; b) $\varphi = -x_1'^2 - x_2'^2, x_1'^2 + x_2'^2 = -1$, a circle of imaginary radius; c) $\varphi = x_1'^2 - x_2'^2, x_1'^2 - x_2'^2 = 1$, an equilateral hyperbola; d) $\varphi = x_1'^2 + 3x_2'^2, x_1'^2 + 3x_2'^2 = 1$, an ellipse with semiaxes 1 and $1/\sqrt{3}$; e) $\varphi = x_1'^2 + \lambda x_2'^2, x_1'^2 + \lambda x_2'^2 = 1$, an ellipse ($\lambda > 0$) or hyperbola ($\lambda < 0$) with semiaxes 1 and $1/\sqrt{\lambda}$ (or $1/\sqrt{-\lambda}$); f) $\varphi = \lambda_1 x_1'^2 + \lambda_2 x_2'^2, \lambda_1 x_1'^2 + \lambda_2 x_2'^2 = 1$, an ellipse if $\lambda_1 > 0, \lambda_2 > 0$, a hyperbola if $\lambda_1\lambda_2 < 0$, an imaginary ellipse if $\lambda_1 < 0, \lambda_2 < 0$.

2. a) $\varphi = x_2'^2, x_2'^2 = 1$, a pair of planes parallel to the $\mathbf{e}_1, \mathbf{e}_3$ plane; b) $\varphi = x_1'^2 + x_2'^2, x_1'^2 + x_2'^2 = 1$, a right circular cylinder; c) $\varphi = x_1'^2 + x_2'^2 - x_3'^2, x_1'^2 + x_2'^2 - x_3'^2 = 1$, a single-sheeted hyperboloid of revolution; d) $\varphi = -x_1'^2 + 2x_2'^2 - x_3'^2, x_1'^2 - 2x_2'^2 + x_3'^2 = -1$, a double-sheeted hyperboloid of revolution; e) $\varphi = a_i a_j x_i x_j$, $a_i a_j x_i x_j = 1$ or $x_1'^2 = c$ after an appropriate coordinate transformation, a pair of parallel planes; f) $\varphi = (a_i a_j + b_i b_j)x_i x_j, (a_i a_j + b_i b_j)x_i x_j = 1$, an elliptic cylinder (cf. Sec. 12, Prob. 13).

3. $\mathbf{u}^* = \mathbf{A}^*\mathbf{x} = x_1\mathbf{e}_1 + (2x_1 + x_2)\mathbf{e}_2 + x_3\mathbf{e}_3$, $\mathbf{A} = \mathbf{A}_1 + \mathbf{A}_2$ where $\mathbf{A}_1\mathbf{x} = (x_1 + x_2)\mathbf{e}_1 + (x_1 + x_2)\mathbf{e}_2 + x_3\mathbf{e}_3$, $\mathbf{A}_2\mathbf{x} = x_2\mathbf{e}_1 - x_1\mathbf{e}_2$; b) $\mathbf{u}^* = \mathbf{A}^*\mathbf{x} = x_2\mathbf{e}_1 - x_1\mathbf{e}_2 + x_3\mathbf{e}_3$, $\mathbf{A} = \mathbf{A}_1 + \mathbf{A}_2$ where $\mathbf{A}_1\mathbf{x} = x_3\mathbf{e}_3$, $\mathbf{A}_2\mathbf{x} = -x_2\mathbf{e}_1 + x_1\mathbf{e}_2$; c) $\mathbf{u}^* = \mathbf{A}^*\mathbf{x} = (\mathbf{b}\cdot\mathbf{x})\mathbf{a}$, $\mathbf{A} = \mathbf{A}_1 + \mathbf{A}_2$ where $\mathbf{A}_1\mathbf{x} = \frac{1}{2}[(\mathbf{a}\cdot\mathbf{x})\mathbf{b} + (\mathbf{b}\cdot\mathbf{x})\mathbf{a}]$, $\mathbf{A}_2\mathbf{x} = \frac{1}{2}[(\mathbf{a}\cdot\mathbf{x})\mathbf{b} - (\mathbf{b}\cdot\mathbf{x})\mathbf{a}]$; d) $\mathbf{u}^* = \mathbf{A}^*\mathbf{x} = (\mathbf{b}_1\cdot\mathbf{x})\mathbf{a}_1 + (\mathbf{b}_2\cdot\mathbf{x})\mathbf{a}_2$, $\mathbf{A} = \mathbf{A}_1 + \mathbf{A}_2$ where $\mathbf{A}_1\mathbf{x} = \frac{1}{2}[(\mathbf{a}_1\cdot\mathbf{x})\mathbf{b}_1 + (\mathbf{b}_1\cdot\mathbf{x})\mathbf{a}_1 + (\mathbf{a}_2\cdot\mathbf{x})\mathbf{b}_2 + (\mathbf{b}_2\cdot\mathbf{x})\mathbf{a}_2]$, $\mathbf{A}_2\mathbf{x} = \frac{1}{2}[(\mathbf{a}_1\cdot\mathbf{x})\mathbf{b}_1 - (\mathbf{b}_1\cdot\mathbf{x})\mathbf{a}_1 + (\mathbf{a}_2\cdot\mathbf{x})\mathbf{b}_2 - (\mathbf{b}_2\cdot\mathbf{x})\mathbf{a}_2]$; e) $\mathbf{u}^* = \mathbf{A}^*\mathbf{x} = \mathbf{x} \times \mathbf{a}$, $\mathbf{A} = \mathbf{A}_1 + \mathbf{A}_2$ where $\mathbf{A}_1 = \mathbf{N}$, $\mathbf{A}_2\mathbf{x} = \mathbf{a} \times \mathbf{x}$.

5. Yes.

7. Write $\mathbf{x}$ in the form $\mathbf{x} = \mathbf{x}_1 + \mathbf{x}_2$, where $\mathbf{x}_1$ is the projection of $\mathbf{x}$ onto Π parallel to l and $\mathbf{x}_2$ is the projection of $\mathbf{x}$ onto l parallel to Π. Then $\mathbf{A}\mathbf{x} = \mathbf{x}_1 - \mathbf{x}_2$. Now show that $(\mathbf{y}, \mathbf{A}\mathbf{x}) = (\mathbf{x}, \mathbf{A}\mathbf{y})$ if and only if l is perpendicular to Π.

8. In c) and d) use integration by parts.

Sec. 16

2. $A = \begin{pmatrix} k & 0 \\ 0 & \frac{1}{k} \end{pmatrix}$.

3. $A = \begin{pmatrix} \cos\alpha & -\frac{a_1}{a_2}\sin\alpha \\ \frac{a_2}{a_1}\sin\alpha & \cos\alpha \end{pmatrix}$

4. Use mathematical induction.

5. $A^n = \begin{pmatrix} \lambda_1^n & 0 & 0 \\ 0 & \lambda_2^n & 0 \\ 0 & 0 & \lambda_3^n \end{pmatrix}$.

8. Use the theorems in Secs. 15.3 and 16.2.

12. Use the theorem in Sec. 14.3.

14. Compare corresponding elements of the matrices AB and BA, using the arbitrariness of B.

15. Same hint.

16. a) $\begin{pmatrix} a & 2b \\ 3b & a + 3b \end{pmatrix}$, where a and b are arbitrary numbers;

b) $\begin{pmatrix} a & b & c \\ 0 & a & b \\ 0 & 0 & a \end{pmatrix}$, where a, b, and c are arbitrary numbers.

17. $\begin{pmatrix} a & b \\ c & -a \end{pmatrix}$, where a, b, and c are arbitrary numbers satisfying the condition $a^2 + bc = 0$.

18. $\pm E$ and $\begin{pmatrix} a & b \\ c & -a \end{pmatrix}$, where a, b, and c are arbitrary numbers satisfying the condition $a^2 + bc = 1$.

20. Both.

22. $A^{n+1}P(t) = P^{(n+1)}(t) = 0,$

$$A^2 = \begin{pmatrix} 0 & 0 & 1\cdot 2 & 0 & \cdots & 0 \\ 0 & 0 & 0 & 2\cdot 3 & \cdots & 0 \\ \cdot & \cdot & \cdot & \cdot & \cdots & \cdot \\ 0 & 0 & 0 & 0 & \cdots & (n-1)n \\ 0 & 0 & 0 & 0 & \cdots & 0 \\ 0 & 0 & 0 & 0 & \cdots & 0 \end{pmatrix},$$

$$A^3 = \begin{pmatrix} 0 & 0 & 0 & 1\cdot 2\cdot 3 & 0 & \cdots & 0 \\ 0 & 0 & 0 & 0 & 2\cdot 3\cdot 4 & \cdots & 0 \\ \cdot & \cdot & \cdot & \cdot & \cdot & \cdots & \cdot \\ 0 & 0 & 0 & 0 & 0 & \cdots & (n-2)(n-1)n \\ 0 & 0 & 0 & 0 & 0 & \cdots & 0 \\ 0 & 0 & 0 & 0 & 0 & \cdots & 0 \\ 0 & 0 & 0 & 0 & 0 & \cdots & 0 \end{pmatrix}, \text{ etc.}$$

The null space of $\mathbf{A}^2, \mathbf{A}^3, \ldots$ is the set of all polynomials of degree not exceeding $1, 2, \ldots$, the range is the set of all polynomials of degreen ot exceeding $1, 2, \ldots$; the defect of $\mathbf{A}^2, \mathbf{A}^3, \ldots$ is $2, 3, \ldots$, the rank $n-1, n-2, \ldots$.

23. The transformation **B** raises the degree of polynomials, and hence can be considered only in the space of all polynomials (of arbitrary degree).

Sec. 17

1. a) $\begin{pmatrix} 5 & -2 \\ -7 & 3 \end{pmatrix}$; b) $\begin{pmatrix} \cos\alpha & \sin\alpha \\ -\sin\alpha & \cos\alpha \end{pmatrix}$; c) $\begin{pmatrix} 1 & -1 & 0 \\ 0 & 1 & -1 \\ 0 & 0 & 1 \end{pmatrix}$;

d) $\begin{pmatrix} 1 & 0 & 0 \\ -a & 1 & 0 \\ a^2 & -a & 1 \end{pmatrix}$; e) $\frac{1}{23}\begin{pmatrix} 4 & 3 & 14 \\ 8 & 6 & 5 \\ -1 & 5 & 8 \end{pmatrix}$.

2. a) $\begin{pmatrix} -10 \\ 19 \end{pmatrix}$; b) $\begin{pmatrix} 3 \\ 2 \\ 2 \end{pmatrix}$; c) $\frac{1}{27}\begin{pmatrix} 30 & 27 \\ -16 & 9 \end{pmatrix}$;

d) $\frac{1}{23}\begin{pmatrix} -20 & -15 & -1 \\ 49 & 77 & 206 \\ 88 & 112 & 239 \end{pmatrix}$.

Sec. 18

1. a), c), e), and f) are groups; b), d), and g) are not.

2. a) Reflection in the diagonals, rotation about the center through 180° and 360°; b) Reflection in the diagonal and in the lines joining midpoints of opposite sides, rotation about the center through 90°, 180°, 270°, and 360°; c) Reflection in the altitudes, rotation about the center through 120°, 240°, and

360°; d) Reflection in the diagonals joining opposite vertices, rotation about the center through 60°, 120°, 180°, 240°, 300° and 360°. The sets of transformations are all groups.

3. All but d) are groups.

4. a), b), d), e), and g) are groups; c) and f) are not.

5. For example, the group in Prob. 3a is a subgroup of the group in Prob. 3b, while the groups in Probs. 3a and 3b are subgroups of the group in Prob. 3c.

7. The diagonal elements must all equal ± 1.

8. a) $E = \begin{pmatrix} 1 & 0 & 0 \\ 0 & 1 & 0 \\ 0 & 0 & 1 \end{pmatrix}, \quad -E = \begin{pmatrix} -1 & 0 & 0 \\ 0 & -1 & 0 \\ 0 & 0 & -1 \end{pmatrix};$

b) $E = \begin{pmatrix} 1 & 0 \\ 0 & 1 \end{pmatrix}, \quad A = \begin{pmatrix} \cos\frac{2\pi}{n} & -\sin\frac{2\pi}{n} \\ \sin\frac{2\pi}{n} & \cos\frac{2\pi}{n} \end{pmatrix};$

c) $E = \begin{pmatrix} 1 & 0 & 0 \\ 0 & 1 & 0 \\ 0 & 0 & 1 \end{pmatrix}, \quad A = \begin{pmatrix} 1 & 0 & 0 \\ 0 & -1 & 0 \\ 0 & 0 & -1 \end{pmatrix}, \quad B = \begin{pmatrix} -1 & 0 & 0 \\ 0 & 1 & 0 \\ 0 & 0 & -1 \end{pmatrix},$

$C = \begin{pmatrix} -1 & 0 & 0 \\ 0 & -1 & 0 \\ 0 & 0 & 0 \end{pmatrix}$, where the coordinate axes are chosen as the axes of rotation.

Chapter 4

Sec. 19

2. a) The eigenvectors are collinear with $\mathbf{b}$, $\lambda = \mathbf{a}\cdot\mathbf{b}$; b) The eigenvectors are collinear with $\mathbf{a}$, $\lambda = 0$; c) $\mathbf{x} = \boldsymbol{\omega}$, $\lambda = 1$; d) $\mathbf{x}_1 = \mathbf{a}\times\mathbf{b}$, $\mathbf{x}_2 = \mathbf{a}+\mathbf{b}$, $\mathbf{x}_3 = \mathbf{a}-\mathbf{b}$, $\lambda_1 = 0$, $\lambda_2 = \mathbf{a}\cdot\mathbf{b}+\mathbf{a}\cdot\mathbf{a}$, $\lambda_3 = -\mathbf{a}\cdot\mathbf{b}+\mathbf{a}\cdot\mathbf{a}$; e) $\mathbf{x}_1 = \mathbf{a}+\mathbf{b}+\mathbf{c}$ is an eigenvector and so is any vector in the plane perpendicular to $\mathbf{x}_1$, $\lambda_1 = \mathbf{a}\cdot\mathbf{a} + 2\mathbf{a}\cdot\mathbf{b}$, $\lambda_2 = \lambda_3 = \mathbf{a}\cdot\mathbf{a} - \mathbf{a}\cdot\mathbf{b}$.

3. a) $\mathbf{x} = (1/\sqrt{3})(\mathbf{e}_1+\mathbf{e}_2+\mathbf{e}_3)$, $\lambda = 1$; b) $\mathbf{x}_1 = (1/\sqrt{3})(\mathbf{e}_1+\mathbf{e}_2+\mathbf{e}_3)$, $\mathbf{x}_2 = (1/\sqrt{6})(\mathbf{e}_1+\mathbf{e}_2-2\mathbf{e}_3)$, $\mathbf{x}_3 = (1/\sqrt{2})(-\mathbf{e}_1+\mathbf{e}_2)$, $\lambda_1 \parallel 2$, $\lambda_2 = \lambda_3 = -1$.

4. a) $\mathbf{x}_1 = (1/\sqrt{2})(\mathbf{e}_1-\mathbf{e}_2)$, $\mathbf{x}_2 = (1/\sqrt{5})(\mathbf{e}_1+2\mathbf{e}_2)$, $\lambda_1 = 1$, $\lambda_2 = 4$; b) $\mathbf{x}_1 = \mathbf{e}_1$, $\mathbf{x}_2 = (1/\sqrt{2})(\mathbf{e}_1+\mathbf{e}_3)$, $\mathbf{x}_3 = (1/\sqrt{2})(\mathbf{e}_2-\mathbf{e}_3)$, $\lambda_1 = 2$, $\lambda_2 = 1$, $\lambda_3 = -1$; c) $\mathbf{x}_1 = (1/\sqrt{3})(\mathbf{e}_1+\mathbf{e}_2-\mathbf{e}_3)$, $\mathbf{x}_2 = (1/\sqrt{2})(\mathbf{e}_1-\mathbf{e}_2)$, $\mathbf{x}_3 = (1/\sqrt{6}) \times (\mathbf{e}_1+\mathbf{e}_2+2\mathbf{e}_3)$, $\lambda_1 = 0$, $\lambda_2 = -1$, $\lambda_3 = 9$; d) $\mathbf{x} = (1/\sqrt{a^4+a^2+1}) \times (a^2\mathbf{e}_1 + a\mathbf{e}_2 + \mathbf{e}_3)$, $\lambda = a$; e) $x = \mathbf{e}_3$, $\lambda = a$; f) $\mathbf{x}_1 = \mathbf{e}_1$, $\mathbf{x}_2 = -b_1\mathbf{e}_1 + (a_1-b_2)\mathbf{e}_2$, $\mathbf{x}_3 = (b_1c_2 - b_2c_1 + c_1c_3)\mathbf{e}_1 + c_2(c_3-a_1)\mathbf{e}_2 + (c_3-b_2)(c_3-a_1)\mathbf{e}_3$, $\lambda_1 = a_1$, $\lambda_2 = b_2$, $\lambda_3 = c_3$.

8. Show that

$$|A^{-1} - \lambda E| = (-\lambda)^n |A^{-1}| \left| A - \frac{1}{\lambda} E \right|$$

if A is of order n.

9. Show that the characteristic equations of the matrices AB and BA have the same coefficients I_1, I_2, I_3.

10. Show that if α is the angle of rotation, then $\lambda_1 = 1$, $\lambda_{2,3} = \cos\alpha \pm i \sin\alpha$. Deduce from this that $2\cos\alpha = a_{ii} - 1$. The direction of l is that of the eigenvector with eigenvalue $\lambda_1 = 1$.

11. $\alpha = \arccos \frac{2}{3}$, l has the direction of the vector $(1/\sqrt{5})(\mathbf{e}_1 + 2\mathbf{e}_2)$.

12. $B = A^{-1} = A^*$.

14. Write the transformation $\mathbf{A}^2 - \mu^2\mathbf{E}$ as a product of two factors.

17. Any number α is an eigenvalue, with $ce^{\alpha t}$ as the corresponding eigenvector.

18. The unique eigenvalue is $\lambda = 0$, with the polynomials of zero degree as the corresponding eigenvectors.

Sec. 20

2. In Example 1 the matrix is $\begin{pmatrix} \lambda & 0 & 0 \\ 0 & \lambda & 0 \\ 0 & 0 & \lambda \end{pmatrix}$ in any basis; in Example 2 the matrix is $\begin{pmatrix} 1 & 0 \\ 0 & \lambda \end{pmatrix}$ in the basis $\mathbf{e}_1, \mathbf{e}_2$; in Example 5 the matrix is $\begin{pmatrix} \lambda_1 & 0 & 0 \\ 0 & \lambda_2 & 0 \\ 0 & 0 & \lambda_3 \end{pmatrix}$ in the basis $\mathbf{e}_1, \mathbf{e}_2, \mathbf{e}_3$. In Prob. 2d the matrix is $\begin{pmatrix} 0 & 0 & 0 \\ 0 & \mathbf{a}\cdot\mathbf{b} + \mathbf{a}\cdot\mathbf{a} & 0 \\ 0 & 0 & -\mathbf{a}\cdot\mathbf{b} + \mathbf{a}\cdot\mathbf{a} \end{pmatrix}$ in the basis $\mathbf{x}_1, \mathbf{x}_2, \mathbf{x}_3$, in Prob. 4a the matrix is $\begin{pmatrix} 1 & 0 \\ 0 & 4 \end{pmatrix}$ in the basis $\mathbf{x}_1, \mathbf{x}_2$; in Prob. 4b the matrix is $\begin{pmatrix} 2 & 0 & 0 \\ 0 & 1 & 0 \\ 0 & 0 & -1 \end{pmatrix}$ in the basis $\mathbf{x}_1, \mathbf{x}_2, \mathbf{x}_3$; in Prob. 4c the matrix is $\begin{pmatrix} 0 & 0 & 0 \\ 0 & -1 & 0 \\ 0 & 0 & 9 \end{pmatrix}$ in the basis $\mathbf{x}_1, \mathbf{x}_2, \mathbf{x}_3$; in Prob. 4f the matrix is $\begin{pmatrix} a_1 & 0 & 0 \\ 0 & b_2 & 0 \\ 0 & 0 & c_3 \end{pmatrix}$ in the basis $\mathbf{x}_1, \mathbf{x}_2, \mathbf{x}_3$.

3. When $\alpha_2^2 \neq \alpha_1\alpha_3$, $\alpha_1\alpha_3 > 0$.

6. Show that the matrix of a proper orthogonal transformation $\mathbf{A}$ is of the form

$$A = \begin{pmatrix} a & -b \\ b & a \end{pmatrix},$$

where $a^2 + b^2 = 1$. Show that an improper orthogonal transformation has real eigenvalues and eigenvectors, and go over to the basis consisting of the eigenvectors.

7. Show that there is always one real eigenvalue equal to ± 1, so that $\mathbf{Ax} = \pm\mathbf{x}$ for the corresponding eigenvector $\mathbf{x}$. Show that the plane perpendicular to $\mathbf{x}$ is invariant under $\mathbf{A}$.

Sec. 21

1. $\begin{pmatrix} 14 & 2 \\ 3 & 14 \end{pmatrix}$.

4. a) Use the fact that $\mathbf{A}^3 = I_1\mathbf{A}^2 - I_2\mathbf{A} + I_3\mathbf{E}$; b) $\mathbf{A}(\alpha\mathbf{a} + \beta\mathbf{a}_1) = \alpha\mathbf{a}_1 + \beta\mathbf{a}_2$, but $\mathbf{a}_2$ is coplanar with the plane of $\mathbf{a}$ and $\mathbf{a}_1$; $\begin{pmatrix} 0 & 0 & I_3 \\ 1 & 0 & -I_2 \\ 0 & 1 & I_1 \end{pmatrix}$.

5. Show that $AB - BA = E$ implies $A^kB - BA^k = kA^{k-1}$, and hence $f(A)B - Bf(A) = f'(A)$ for any polynomial $f(\lambda)$. But this cannot hold for the polynormial $g(\lambda)$ of minimum degree for which $g(A) = 0$.

Sec. 22

2. Choose the eigenvectors as a basis, and note that a symmetric matrix A goes into a symmetric matrix A' under an orthogonal transformation Γ, since

$$(A')^* = (\Gamma A \Gamma^{-1})^* = (\Gamma^{-1})^* A^* \Gamma^* = (\Gamma^{-1})^* A \Gamma^* = \Gamma A \Gamma^{-1} = A'$$

(see Secs. 16.4 and 18.3).

3. Show that the subspace consisting of all eigenvectors of one transformation corresponding to the same eigenvalue is invariant under the other transformation.

4. Proved by analogy with the corresponding properties of a symmetric linear transformation.

Sec. 23

1. a) $\frac{1}{\sqrt{5}}(\mathbf{e}_1 - 2\mathbf{e}_2), \frac{1}{\sqrt{5}}(2\mathbf{e}_1 + \mathbf{e}_2), \begin{pmatrix} 2 & 0 \\ 0 & 7 \end{pmatrix}$; b) $\frac{1}{3}(\mathbf{e}_1 + 2\mathbf{e}_2 + 2\mathbf{e}_3)$, $\frac{1}{3}(2\mathbf{e}_1 + \mathbf{e}_2 - 2\mathbf{e}_3), \frac{1}{3}(-2\mathbf{e}_1 + 2\mathbf{e}_2 - \mathbf{e}_3), \begin{pmatrix} 3 & 0 & 0 \\ 0 & 6 & 0 \\ 0 & 0 & 9 \end{pmatrix}$; c) $\frac{1}{3}(2\mathbf{e}_1 + \mathbf{e}_2 - 2\mathbf{e}_3)$, $\frac{1}{3}(\mathbf{e}_1 + 2\mathbf{e}_2 + 2\mathbf{e}_3), \frac{1}{3}(2\mathbf{e}_1 - 2\mathbf{e}_2 + \mathbf{e}_3), \begin{pmatrix} 6 & 0 & 0 \\ 0 & -3 & 0 \\ 0 & 0 & -3 \end{pmatrix}$; d) $\frac{1}{\sqrt{2}}(\mathbf{e}_1 - \mathbf{e}_3)$, $\frac{1}{\sqrt{2}}(\mathbf{e}_1 + \mathbf{e}_3), \mathbf{e}_2, \begin{pmatrix} -1 & 0 & 0 \\ 0 & 1 & 0 \\ 0 & 0 & 1 \end{pmatrix}$.

2. a) $\frac{1}{5}\begin{pmatrix} 2^{30}+4\cdot 7^{30} & -2^{31}+2\cdot 7^{30} \\ -2^{31}+2\cdot 7^{30} & 2^{32}+7^{30} \end{pmatrix}$;

b) $3^{38}\begin{pmatrix} 1+2^{32}+4\cdot 3^{30} & 2+2^{31}-4\cdot 3^{30} & 2-2^{32}+2\cdot 3^{30} \\ 2+2^{31}-4\cdot 3^{30} & 4+2^{30}+4\cdot 3^{30} & 4-2^{31}-2\cdot 3^{30} \\ 2-2^{32}+2\cdot 3^{30} & 4-2^{31}-2\cdot 3^{30} & 4+2^{32}+3^{30} \end{pmatrix}$.

Hint. Reduce the matrices to diagonal form (see Probs. la, lb), raise the diagonal matrices to the thirtieth power, and then transform back to the old basis.

3. a) To prove the necessity, apply **A** to the eigenvectors, while to prove the sufficiency, consider the basis consisting of the eigenvectors; b) Let **B** be such that $\mathbf{B}\mathbf{e}_i = \sqrt{\lambda_i}\mathbf{e}_i$ (no summation over i), where $\mathbf{e}_1, \mathbf{e}_2, \mathbf{e}_3$ is the basis consisting of the eigenvectors of **A**; c) Show that if all the λ_i are distinct, then the matrix C is diagonal, while if $\lambda_1 = \lambda_2 \neq \lambda_3$, then $c_{13} = c_{31} = c_{23} = c_{32} = 0$; in each case (including $\lambda_1 = \lambda_2 = \lambda_3$), verify the formula $BC = CB$ directly; d) $(\mathbf{x}, (\mathbf{A}+\mathbf{B})\mathbf{x}) = (\mathbf{x}, \mathbf{A}\mathbf{x}) + (\mathbf{x}, \mathbf{B}\mathbf{x}) \geq 0$, $(\mathbf{A}+\mathbf{B})^* = \mathbf{A}^* + \mathbf{B}^* = \mathbf{A}+\mathbf{B}$; e) Let $\mathbf{A}_1^2 = \mathbf{A}$, $\mathbf{B}_1^2 = \mathbf{B}$, $\mathbf{C} = \mathbf{A}_1\mathbf{B}_1$; show that $\mathbf{A}_1\mathbf{B}_1 = \mathbf{B}_1\mathbf{A}_1$, $\mathbf{C}^2 = \mathbf{AB}$ and hence that **AB** is nonnegative (the symmetry follows from Sec. 16, Prob. 6); f) Use parts d) and e); g) Use the result of Sec. 19, Prob. 6.

4. a) $B = \frac{1}{3}A$; b) $B = \begin{pmatrix} 3 & 2 & 0 \\ 2 & 4 & 2 \\ 0 & 2 & 5 \end{pmatrix}$.

5. Show that the eigenvalues of an orthogonal symmetric transformation equal ± 1.

Sec. 24

1. a) $\varphi = \frac{1}{2}(x_{1'}^2 - x_{2'}^2)$; b) $\varphi = 2x_{1'}^2$; c) $\varphi = \frac{1}{2}(3x_{1'}^2 + x_{2'}^2)$; d) $\varphi = 4x_{1'}^2 + 4x_{2'}^2 - 2x_{3'}^2$; e) $\varphi = x_{1'}^2 + \sqrt{3}x_{2'}^2 - \sqrt{3}x_{3'}^2$.

2. a) $\varphi = x_{1'}^2 + 9x_{2'}^2$, $\Gamma = \frac{1}{\sqrt{2}}\begin{pmatrix} -1 & 1 \\ 1 & 1 \end{pmatrix}$; b) $\varphi = x_{1'}^2 - \frac{1}{2}x_{2'}^2 - \frac{1}{2}x_{3'}^2$,

$\Gamma = \frac{1}{\sqrt{6}}\begin{pmatrix} \sqrt{2} & \sqrt{2} & \sqrt{2} \\ \sqrt{3} & -\sqrt{3} & 0 \\ 1 & 1 & -2 \end{pmatrix}$; c) $\varphi = 3x_{1'}^2 + 6x_{2'}^2 + 9x_{3'}^2$,

$\Gamma = \frac{1}{3}\begin{pmatrix} 1 & 2 & 2 \\ 2 & 1 & -2 \\ 2 & -2 & 1 \end{pmatrix}$; d) $\varphi = 4x_{1'}^2 + x_{2'}^2 - 2x_{3'}^2$, $\Gamma = \frac{1}{3}\begin{pmatrix} 2 & -2 & 1 \\ 2 & 1 & -2 \\ 1 & 2 & 2 \end{pmatrix}$;

e) $\varphi = 7x_{1'}^2 - 2x_{2'}^2 + 7x_{3'}^2$, $\Gamma = \frac{1}{6}\begin{pmatrix} 3\sqrt{2} & 0 & -3\sqrt{2} \\ 4 & 2 & 4 \\ \sqrt{2} & -4\sqrt{2} & \sqrt{2} \end{pmatrix}$.

3. a) $a > \frac{1}{3}$; $a > 2$; c) $|a| < \sqrt{\frac{5}{3}}$.

4. In the basis consisting of the unit eigenvectors $\mathbf{e}_{1'}$ and $\mathbf{e}_{2'}$, we have

$$\lambda_1(x_{1'}^2 + x_{2'}^2) \leq \lambda_1 x_{1'}^2 + \lambda_2 x_{2'}^2 \leq \lambda_2(x_{1'}^2 + x_{2'}^2).$$

Now use the invariance of $x_1^2 + x_2^2$.

5. Show that the eigenvalues of the matrix $A - xE$ are obtained by subtracting x from the eigenvalues of the matrix A.

6. a) A single-sheeted hyperboloid of revolution with axis $\mathbf{e}_{3'}$; b) A double-sheeted hyperboloid of revolution with axis $\mathbf{e}_{3'}$; c) An ellipsoid; d) A single-sheeted hyperboloid; e) A double-sheeted hyperboloid; f) An "imaginary" ellipsoid.

Sec. 25

1. Start from the transformation **AA***.

2. a) $$A = \begin{pmatrix} \frac{\sqrt{3}}{2} & -\frac{1}{2} \\ \frac{1}{2} & \frac{\sqrt{3}}{2} \end{pmatrix} \begin{pmatrix} \frac{4+\sqrt{3}}{2} & -\frac{1}{2} \\ -\frac{1}{2} & \frac{4-\sqrt{3}}{2} \end{pmatrix};$$

b) $$A = \begin{pmatrix} \frac{1}{\sqrt{2}} & -\frac{1}{\sqrt{2}} \\ \frac{1}{\sqrt{2}} & \frac{1}{\sqrt{2}} \end{pmatrix} \begin{pmatrix} \sqrt{2} & 0 \\ 0 & 4\sqrt{2} \end{pmatrix}.$$

3. $$A = \begin{pmatrix} \frac{14}{3} & \frac{2}{3} & -\frac{4}{3} \\ \frac{2}{3} & \frac{17}{3} & \frac{2}{3} \\ -\frac{4}{3} & \frac{2}{3} & \frac{14}{3} \end{pmatrix} \begin{pmatrix} \frac{2}{3} & -\frac{1}{3} & \frac{2}{3} \\ \frac{2}{3} & \frac{2}{3} & -\frac{1}{3} \\ -\frac{1}{3} & \frac{2}{3} & \frac{2}{3} \end{pmatrix}.$$

BIBLIOGRAPHY

Borisenko, A. I. and I. E. Tarapov, *Vector and Tensor Analysis with Applications*, translated by R. A. Silverman, Prentice-Hall (1968).

Hoffman, K. and R. Kunze, *Linear Algebra*, Prentice-Hall (1961).

Noble, B., *Applied Linear Algebra*, Prentice-Hall (1969).

Shilov, G. E., *An Introduction to the Theory of Linear Spaces*, translated by R. A. Silverman, Prentice-Hall (1961).

Shilov, G. E., *Linear Algebra*, translated by R. A. Silverman, Prentice-Hall (1971).

INDEX

A

Adjoint (transformation), 80
Antisymmetric Kronecker symbol, 17

B

Basis, 9
  left-handed, 12
  orthonormal, 12
  right-handed, 12
Basis transformation, 22 ff.
  inverse, 23
  matrix of, 23
Bessel's inequality, 16
Bilinear form, 41 ff.
  antisymmetric, 56
  antisymmetrization of, 57
  coefficients of, 42
    transformation law of, 43
  matrix of, 42
  polar, 59
  symmetric, 55, 59
  symmetrization of, 57
Bilinear function (*see* Bilinear form)
Borisenko, A. I., 18

C

Canonical basis, 135
Canonical form, 134
Cauchy-Schwarz inequality, 14
Center (of a curve or surface), 35
Characteristic equation, 110
  of a symmetric transformation, 125
Characteristic polynomial, 110
Characteristic surface, 60–62, 82
Contraction of tensors, 51–53
Cramer's theorem, 21
Cross product (*see* Vector product)
Cubic form, 60

D

Defect, 78
Diagonal matrix, 73
Diagonalization of a symmetric transformation, 127–132
Distance:
  between two points, 31
  from a point to a plane, 33
Division of a line segment, 31
Dimension, 8

E

Eigenvalue(s), 107 ff.
  of a symmetric transformation, 125
Eigenvector(s), 107 ff.
  of a symmetric transformation, 124–126
Einstein, A., 10
Equation of a straight line, 33
  parametric equations of, 33
  vector form of, 33
Euclidean space, 14
Eulerian angles, 29
Expansion (contraction), 65, 66, 72

F

Full linear group:
 of dimension n, 99
 of dimension two, 99
 of dimension three, 99

G

Group, 99
 commutative, 104
 cyclic, 104
 full linear, 99
 multiplication table of, 103
 of orthogonal transformations, 101
 symmetry, 103
 unimodular, 100

H

Hamilton-Cayley theorem, 122
Homogeneous expansion (contraction), 65
Homothetic transformation, 65

I

Identity transformation, 65, 99
Image, 78
Index of summation, 10
Invariance, 25
Invariant, 47, 111
Invariant plane, 116
Inverse image, 78
Inverse transformation, 94, 99

J

Jacobi's identity, 21

K

Kronecker delta, 12
Kronecker symbol:
 antisymmetric, 17
 symmetric, 12

L

Lagrange's identity, 21
Legendre polynomials, 16
Level surfaces, 40
Line of nodes, 29
Linear combination, 4
Linear dependence, 4
Linear form, 38
 coefficients of, 39
 transformation law of, 39
Linear function (*see* Linear form)
Linear independence, 4
Linear operator (*see* Linear transformation)
Linear space(s), 1 ff.
 basis for, 9
 dimension of, 8
Linear subspace(s), 3–4
 intersection of, 4
 spanned by given vectors, 11
 sum of, 4
 trivial, 3
Linear transformation(s), 64 ff.
 adjoint of, 80
 antisymmetric, 82–83
 as a group, 98–105
 characteristic equation of, 110
 commuting, 88
 defect of, 78
 diagonalization of, 118
 eigenvalue of, 107
 eigenvector of, 107
 invariants of, 111
 inverse of, 94–97, 99
 matrix of, 69
 multiplication of, 87
  associativity of, 99
  noncommutativity of, 88
 nonsingular, representation of, 138–143
 null space of, 78

Linear transformation(s) (*cont.*):
  orthogonal, 101
  polynomial in a, 121
  product of, 87
  product of, with a number, 80
  range of, 78
  rank of, 78
  self-adjoint, 81
  singular, 75
  sum of, 79
  symmetric, 81
    diagonalization of, 127–132
    eigenvalues of, 125
    eigenvectors of, 124–126
    nonnegative, 133
Linear vector function (*see* Linear transformation)

M

Magnification, 74
Matrix, 23 ff.
  (descending) principal minors of, 135
  diagonal, 73
  idempotent, 94
  inverse, 95
  involutory, 94
  null, 71
  of a basis transformation, 23
    determinant of, 24
  of a linear transformation, 69
    determinant of, 74
  order of, 23
  orthogonal, 24, 102
    determinant of, 24, 102
  rank of, 75
  singular, 75
  skew-symmetric, 57
  symmetric, 55
  trace of, 93
  transposition of, 80
  triangular, 98
  unit, 71, 89
  zero, 71
Matrix polynomial, 121
  root of, 122
Matrix product, 88–89
  determinant of, 90
Menelaus:
  direct theorem of, 21
  indirect theorem of, 21
Multilinear form, 44
  antisymmetric, 57
  degree of, 44
  symmetric, 56
Multilinear function (*see* Multilinear form)

N

Null matrix, 71
Null space, 78
Null transformation, 65

O

Origin of coordinates, 1
Orthogonal matrix, 24
  determinant of, 24
Orthogonal transformation, 101
  improper, 103
  proper, 103
Orthogonality relations, 23, 102

P

Parseval's theorem, 16
Plane:
  distance from a point to a, 33
  equation of, 32
    vector form of, 32
p-linear form, 44
Polar form, 59
Preimage (*see* Inverse image)
Principal directions, 134
Principal minors, 135
Projection, 14, 66

Q

Quadratic form, 59, 134–137
negative definite, 135
positive definite, 135
principal directions of, 134
reduction of, to canonical form, 134

R

Radius vector, 30
Range, 78
Rank, 75, 78
Rotation, 66

S

Scalar function, 38
Scalar product, 12
invariance of, 26–27
Scalar quantity, 47
Scalar triple product, 18
Second-degree curve, 34
determination of center of, 35–36
Second-degree surface, 34–35
determination of center of, 35–36
Secular equation (*see* Characteristic equation)
Shilov, G. E., 8, 123
Silverman, R. A., 8, 18, 82, 95, 109
Square matrix, 23
Straight line:
as intersection of two planes, 33–34
equation of, 33
vector form of, 33
parametric equations of, 33
Subgroup, 100
finite, 103
Summation convention, 10
Sylvester's criterion, 135
Symmetric transformation, 81, 124 ff.
characteristic equation of, 125
Symmetric transformation (*cont.*):
diagonalization of, 127–132
eigenvalues of, 125
eigenvectors of, 124–126
nonnegative, 133

T

Tarapov, I. E., 18
Tensor(s), 46 ff.
antisymmetric, 57
antisymmetrization of, 58
characteristic surface of, 60–62, 82
components of, 46
contraction of, 51–53
determined by a form, 46
discriminantal, 47
null, 47
of order *p*, 46
of order zero, 47
orthogonal, 46
permutation of indices of, 54
product of, 51
product of, with a real number, 50
spherical, 84
sum of, 50
symmetric, 55, 56, 82
symmetrization of, 58
trace of, 52
transformation law of, 48–49
unit, 47
Tensor calculus, fundamental problem of, 25
Tensor character, test for, 53–54
Trace, 52, 93
Transposition, 80, 90
Triangle inequalities, 14

U

Unit matrix, 71, 89
Unit transformation (*see* Identity transformation)

V

Vector(s), 1 ff.
  angle between, 13, 15
  axial, 18
  closed set of, 1
  collinear, 5
  components of, 9
    rectangular, 12
  coplanar, 5
  free, 1
  length of, 13, 14
  linear combination of, 4
    coefficients of, 4
  linearly dependent, 4
  linearly independent, 4
  negative, 1
  normalized, 13
  orthogonal, 15
  product of, with a number, 1
Vector(s) (*cont.*):
  projection of, 14, 66
  scalar product of, 12
  scalar triple product of, 18
  sum of, 1
  unit, 12
  vector product of, 16
  vector triple product of, 19
  zero, 1
Vector product, 16
  invariance of, 27–28
Vector space (*see* Linear space)
Vector triple product, 19

Z

Zero matrix (*see* Null matrix)
Zero transformation (*see* Null transformation)

A CATALOGUE OF SELECTED DOVER BOOKS IN ALL FIELDS OF INTEREST

A CATALOGUE OF SELECTED DOVER BOOKS IN ALL FIELDS OF INTEREST

THE CURVES OF LIFE, Theodore A. Cook. Examination of shells, leaves, horns, human body, art, etc., in "*the* classic reference on how the golden ratio applies to spirals and helices in nature . . . ."—Martin Gardner. 426 illustrations. Total of 512pp. 5⅜ x 8½. 23701-X Pa. $5.95

AN ILLUSTRATED FLORA OF THE NORTHERN UNITED STATES AND CANADA, Nathaniel L. Britton, Addison Brown. Encyclopedic work covers 4666 species, ferns on up. Everything. Full botanical information, illustration for each. This earlier edition is preferred by many to more recent revisions. 1913 edition. Over 4000 illustrations, total of 2087pp. 6⅛ x 9¼. 22642-5, 22643-3, 22644-1 Pa., Three-vol. set $24.00

MANUAL OF THE GRASSES OF THE UNITED STATES, A. S. Hitchcock, U.S. Dept. of Agriculture. The basic study of American grasses, both indigenous and escapes, cultivated and wild. Over 1400 species. Full descriptions, information. Over 1100 maps, illustrations. Total of 1051pp. 5⅜ x 8½. 22717-0, 22718-9 Pa., Two-vol. set $15.00

THE CACTACEAE,, Nathaniel L. Britton, John N. Rose. Exhaustive, definitive. Every cactus in the world. Full botanical descriptions. Thorough statement of nomenclatures, habitat, detailed finding keys. The one book needed by every cactus enthusiast. Over 1275 illustrations. Total of 1080pp. 8 x 10¼. 21191-6, 21192-4 Clothbd., Two-vol. set $35.00

AMERICAN MEDICINAL PLANTS, Charles F. Millspaugh. Full descriptions, 180 plants covered: history; physical description; methods of preparation with all chemical constituents extracted; all claimed curative or adverse effects. 180 full-page plates. Classification table. 804pp. 6½ x 9¼. 23034-1 Pa. $10.00

A MODERN HERBAL, Margaret Grieve. Much the fullest, most exact, most useful compilation of herbal material. Gigantic alphabetical encyclopedia, from aconite to zedoary, gives botanical information, medical properties, folklore, economic uses, and much else. Indispensable to serious reader. 161 illustrations. 888pp. 6½ x 9¼. (Available in U.S. only) 22798-7, 22799-5 Pa., Two-vol. set $12.00

THE HERBAL or GENERAL HISTORY OF PLANTS, John Gerard. The 1633 edition revised and enlarged by Thomas Johnson. Containing almost 2850 plant descriptions and 2705 superb illustrations, Gerard's *Herbal* is a monumental work, the book all modern English herbals are derived from, the one herbal every serious enthusiast should have in its entirety. Original editions are worth perhaps $750. 1678pp. 8½ x 12¼. 23147-X Clothbd. $50.00

MANUAL OF THE TREES OF NORTH AMERICA, Charles S. Sargent. The basic survey of every native tree and tree-like shrub, 717 species in all. Extremely full descriptions, information on habitat, growth, locales, economics, etc. Necessary to every serious tree lover. Over 100 finding keys. 783 illustrations. Total of 986pp. 5⅜ x 8½. 20277-1, 20278-X Pa., Two-vol. set $10.00

THE DEPRESSION YEARS AS PHOTOGRAPHED BY ARTHUR ROTHSTEIN, Arthur Rothstein. First collection devoted entirely to the work of outstanding 1930s photographer: famous dust storm photo, ragged children, unemployed, etc. 120 photographs. Captions. 119pp. 9¼ x 10¾.
23590-4 Pa. $5.00

CAMERA WORK: A PICTORIAL GUIDE, Alfred Stieglitz. All 559 illustrations and plates from the most important periodical in the history of art photography, *Camera Work* (1903-17). Presented four to a page, reduced in size but still clear, in strict chronological order, with complete captions. Three indexes. Glossary. Bibliography. 176pp. 8⅜ x 11¼.
23591-2 Pa. $6.95

ALVIN LANGDON COBURN, PHOTOGRAPHER, Alvin L. Coburn. Revealing autobiography by one of greatest photographers of 20th century gives insider's version of Photo-Secession, plus comments on his own work. 77 photographs by Coburn. Edited by Helmut and Alison Gernsheim. 160pp. 8⅛ x 11. 23685-4 Pa. $6.00

NEW YORK IN THE FORTIES, Andreas Feininger. 162 brilliant photographs by the well-known photographer, formerly with *Life* magazine, show commuters, shoppers, Times Square at night, Harlem nightclub, Lower East Side, etc. Introduction and full captions by John von Hartz. 181pp. 9¼ x 10¾. 23585-8 Pa. $6.00

GREAT NEWS PHOTOS AND THE STORIES BEHIND THEM, John Faber. Dramatic volume of 140 great news photos, 1855 through 1976, and revealing stories behind them, with both historical and technical information. Hindenburg disaster, shooting of Oswald, nomination of Jimmy Carter, etc. 160pp. 8¼ x 11. 23667-6 Pa. $5.00

THE ART OF THE CINEMATOGRAPHER, Leonard Maltin. Survey of American cinematography history and anecdotal interviews with 5 masters—Arthur Miller, Hal Mohr, Hal Rosson, Lucien Ballard, and Conrad Hall. Very large selection of behind-the-scenes production photos. 105 photographs. Filmographies. Index. Originally *Behind the Camera.* 144pp. 8¼ x 11. 23686-2 Pa. $5.00

DESIGNS FOR THE THREE-CORNERED HAT (LE TRICORNE), Pablo Picasso. 32 fabulously rare drawings—including 31 color illustrations of costumes and accessories—for 1919 production of famous ballet. Edited by Parmenia Migel, who has written new introduction. 48pp. 9⅜ x 12¼. (Available in U.S. only) 23709-5 Pa. $5.00

NOTES OF A FILM DIRECTOR, Sergei Eisenstein. Greatest Russian filmmaker explains montage, making of *Alexander Nevsky,* aesthetics; comments on self, associates, great rivals (Chaplin), similar material. 78 illustrations. 240pp. 5⅜ x 8½. 22392-2 Pa. $4.50

A MAYA GRAMMAR, Alfred M. Tozzer. Practical, useful English-language grammar by the Harvard anthropologist who was one of the three greatest American scholars in the area of Maya culture. Phonetics, grammatical processes, syntax, more. 301pp. 5⅜ x 8½. 23465-7 Pa. $4.00

THE JOURNAL OF HENRY D. THOREAU, edited by Bradford Torrey, F. H. Allen. Complete reprinting of 14 volumes, 1837-61, over two million words; the sourcebooks for *Walden*, etc. Definitive. All original sketches, plus 75 photographs. Introduction by Walter Harding. Total of 1804pp. 8½ x 12¼. 20312-3, 20313-1 Clothbd., Two-vol. set $50.00

CLASSIC GHOST STORIES, Charles Dickens and others. 18 wonderful stories you've wanted to reread: "The Monkey's Paw," "The House and the Brain," "The Upper Berth," "The Signalman," "Dracula's Guest," "The Tapestried Chamber," etc. Dickens, Scott, Mary Shelley, Stoker, etc. 330pp. 5⅜ x 8½. 20735-8 Pa. $3.50

SEVEN SCIENCE FICTION NOVELS, H. G. Wells. Full novels. *First Men in the Moon, Island of Dr. Moreau, War of the Worlds, Food of the Gods, Invisible Man, Time Machine, In the Days of the Comet*. A basic science-fiction library. 1015pp. 5⅜ x 8½. (Available in U.S. only) 20264-X Clothbd. $8.95

ARMADALE, Wilkie Collins. Third great mystery novel by the author of *The Woman in White* and *The Moonstone*. Ingeniously plotted narrative shows an exceptional command of character, incident and mood. Original magazine version with 40 illustrations. 597pp. 5⅜ x 8½. 23429-0 Pa. $5.00

MASTERS OF MYSTERY, H. Douglas Thomson. The first book in English (1931) devoted to history and aesthetics of detective story. Poe, Doyle, LeFanu, Dickens, many others, up to 1930. New introduction and notes by E. F. Bleiler. 288pp. 5⅜ x 8½. (Available in U.S. only) 23606-4 Pa. $4.00

FLATLAND, E. A. Abbott. Science-fiction classic explores life of 2-D being in 3-D world. Read also as introduction to thought about hyperspace. Introduction by Banesh Hoffmann. 16 illustrations. 103pp. 5⅜ x 8½. 20001-9 Pa. $1.75

THREE SUPERNATURAL NOVELS OF THE VICTORIAN PERIOD, edited, with an introduction, by E. F. Bleiler. Reprinted complete and unabridged, three great classics of the supernatural: *The Haunted Hotel* by Wilkie Collins, *The Haunted House at Latchford* by Mrs. J. H. Riddell, and *The Lost Stradivarious* by J. Meade Falkner. 325pp. 5⅜ x 8½. 22571-2 Pa. $4.00

AYESHA: THE RETURN OF "SHE," H. Rider Haggard. Virtuoso sequel featuring the great mythic creation, Ayesha, in an adventure that is fully as good as the first book, *She*. Original magazine version, with 47 original illustrations by Maurice Greiffenhagen. 189pp. 6½ x 9¼. 23649-8 Pa. $3.50

ART FORMS IN NATURE, Ernst Haeckel. Multitude of strangely beautiful natural forms: Radiolaria, Foraminifera, jellyfishes, fungi, turtles, bats, etc. All 100 plates of the 19th-century evolutionist's *Kunstformen der Natur* (1904). 100pp. 9⅜ x 12¼. 22987-4 Pa. $4.50

CHILDREN: A PICTORIAL ARCHIVE FROM NINETEENTH-CENTURY SOURCES, edited by Carol Belanger Grafton. 242 rare, copyright-free wood engravings for artists and designers. Widest such selection available. All illustrations in line. 119pp. 8⅜ x 11¼.
23694-3 Pa. $3.50

WOMEN: A PICTORIAL ARCHIVE FROM NINETEENTH-CENTURY SOURCES, edited by Jim Harter. 391 copyright-free wood engravings for artists and designers selected from rare periodicals. Most extensive such collection available. All illustrations in line. 128pp. 9 x 12.
23703-6 Pa. $4.50

ARABIC ART IN COLOR, Prisse d'Avennes. From the greatest ornamentalists of all time—50 plates in color, rarely seen outside the Near East, rich in suggestion and stimulus. Includes 4 plates on covers. 46pp. 9⅜ x 12¼. 23658-7 Pa. $6.00

AUTHENTIC ALGERIAN CARPET DESIGNS AND MOTIFS, edited by June Beveridge. Algerian carpets are world famous. Dozens of geometrical motifs are charted on grids, color-coded, for weavers, needleworkers, craftsmen, designers. 53 illustrations plus 4 in color. 48pp. 8¼ x 11. (Available in U.S. only) 23650-1 Pa. $1.75

DICTIONARY OF AMERICAN PORTRAITS, edited by Hayward and Blanche Cirker. 4000 important Americans, earliest times to 1905, mostly in clear line. Politicians, writers, soldiers, scientists, inventors, industrialists, Indians, Blacks, women, outlaws, etc. Identificatory information. 756pp. 9¼ x 12¾. 21823-6 Clothbd. $40.00

HOW THE OTHER HALF LIVES, Jacob A. Riis. Journalistic record of filth, degradation, upward drive in New York immigrant slums, shops, around 1900. New edition includes 100 original Riis photos, monuments of early photography. 233pp. 10 x 7⅞. 22012-5 Pa. $6.00

NEW YORK IN THE THIRTIES, Berenice Abbott. Noted photographer's fascinating study of city shows new buildings that have become famous and old sights that have disappeared forever. Insightful commentary. 97 photographs. 97pp. 11⅜ x 10. 22967-X Pa. $5.00

MEN AT WORK, Lewis W. Hine. Famous photographic studies of construction workers, railroad men, factory workers and coal miners. New supplement of 18 photos on Empire State building construction. New introduction by Jonathan L. Doherty. Total of 69 photos. 63pp. 8 x 10¾.
23475-4 Pa. $3.00

THE ANATOMY OF THE HORSE, George Stubbs. Often considered the great masterpiece of animal anatomy. Full reproduction of 1766 edition, plus prospectus; original text and modernized text. 36 plates. Introduction by Eleanor Garvey. 121pp. 11 x 14¾. 23402-9 Pa. $6.00

BRIDGMAN'S LIFE DRAWING, George B. Bridgman. More than 500 illustrative drawings and text teach you to abstract the body into its major masses, use light and shade, proportion; as well as specific areas of anatomy, of which Bridgman is master. 192pp. 6½ x 9¼. (Available in U.S. only) 22710-3 Pa. $3.00

ART NOUVEAU DESIGNS IN COLOR, Alphonse Mucha, Maurice Verneuil, Georges Auriol. Full-color reproduction of *Combinaisons ornementales* (c. 1900) by Art Nouveau masters. Floral, animal, geometric, interlacings, swashes—borders, frames, spots—all incredibly beautiful. 60 plates, hundreds of designs. 9⅜ x 8-1/16. 22885-1 Pa. $4.00

FULL-COLOR FLORAL DESIGNS IN THE ART NOUVEAU STYLE, E. A. Seguy. 166 motifs, on 40 plates, from *Les fleurs et leurs applications decoratives* (1902): borders, circular designs, repeats, allovers, "spots." All in authentic Art Nouveau colors. 48pp. 9⅜ x 12¼. 23439-8 Pa. $5.00

A DIDEROT PICTORIAL ENCYCLOPEDIA OF TRADES AND INDUSTRY, edited by Charles C. Gillispie. 485 most interesting plates from the great French Encyclopedia of the 18th century show hundreds of working figures, artifacts, process, land and cityscapes; glassmaking, papermaking, metal extraction, construction, weaving, making furniture, clothing, wigs, dozens of other activities. Plates fully explained. 920pp. 9 x 12. 22284-5, 22285-3 Clothbd., Two-vol. set $40.00

HANDBOOK OF EARLY ADVERTISING ART, Clarence P. Hornung. Largest collection of copyright-free early and antique advertising art ever compiled. Over 6,000 illustrations, from Franklin's time to the 1890's for special effects, novelty. Valuable source, almost inexhaustible.
Pictorial Volume. Agriculture, the zodiac, animals, autos, birds, Christmas, fire engines, flowers, trees, musical instruments, ships, games and sports, much more. Arranged by subject matter and use. 237 plates. 288pp. 9 x 12. 20122-8 Clothbd. $13.50

Typographical Volume. Roman and Gothic faces ranging from 10 point to 300 point, "Barnum," German and Old English faces, script, logotypes, scrolls and flourishes, 1115 ornamental initials, 67 complete alphabets, more. 310 plates. 320pp. 9 x 12. 20123-6 Clothbd. $15.00

CALLIGRAPHY (CALLIGRAPHIA LATINA), J. G. Schwandner. High point of 18th-century ornamental calligraphy. Very ornate initials, scrolls, borders, cherubs, birds, lettered examples. 172pp. 9 x 13. 20475-8 Pa. $6.00

THE COMPLETE WOODCUTS OF ALBRECHT DURER, edited by Dr. W. Kurth. 346 in all: "Old Testament," "St. Jerome," "Passion," "Life of Virgin," Apocalypse," many others. Introduction by Campbell Dodgson. 285pp. 8½ x 12¼. 21097-9 Pa. $7.50

DRAWINGS OF ALBRECHT DURER, edited by Heinrich Wolfflin. 81 plates show development from youth to full style. Many favorites; many new. Introduction by Alfred Werner. 96pp. 8⅛ x 11. 22352-3 Pa. $5.00

THE HUMAN FIGURE, Albrecht Dürer. Experiments in various techniques—stereometric, progressive proportional, and others. Also life studies that rank among finest ever done. Complete reprinting of *Dresden Sketchbook*. 170 plates. 355pp. 8⅜ x 11¼. 21042-1 Pa. $7.95

OF THE JUST SHAPING OF LETTERS, Albrecht Dürer. Renaissance artist explains design of Roman majuscules by geometry, also Gothic lower and capitals. Grolier Club edition. 43pp. 7⅞ x 10¾ 21306-4 Pa. $8.00

TEN BOOKS ON ARCHITECTURE, Vitruvius. The most important book ever written on architecture. Early Roman aesthetics, technology, classical orders, site selection, all other aspects. Stands behind everything since. Morgan translation. 331pp. 5⅜ x 8½. 20645-9 Pa. $4.00

THE FOUR BOOKS OF ARCHITECTURE, Andrea Palladio. 16th-century classic responsible for Palladian movement and style. Covers classical architectural remains, Renaissance revivals, classical orders, etc. 1738 Ware English edition. Introduction by A. Placzek. 216 plates. 110pp. of text. 9½ x 12¾. 21308-0 Pa. $8.95

HORIZONS, Norman Bel Geddes. Great industrialist stage designer, "father of streamlining," on application of aesthetics to transportation, amusement, architecture, etc. 1932 prophetic account; function, theory, specific projects. 222 illustrations. 312pp. 7⅞ x 10¾. 23514-9 Pa. $6.95

FRANK LLOYD WRIGHT'S FALLINGWATER, Donald Hoffmann. Full, illustrated story of conception and building of Wright's masterwork at Bear Run, Pa. 100 photographs of site, construction, and details of completed structure. 112pp. 9¼ x 10. 23671-4 Pa. $5.50

THE ELEMENTS OF DRAWING, John Ruskin. Timeless classic by great Viltorian; starts with basic ideas, works through more difficult. Many practical exercises. 48 illustrations. Introduction by Lawrence Campbell. 228pp. 5⅜ x 8½. 22730-8 Pa. $2.75

GIST OF ART, John Sloan. Greatest modern American teacher, Art Students League, offers innumerable hints, instructions, guided comments to help you in painting. Not a formal course. 46 illustrations. Introduction by Helen Sloan. 200pp. 5⅜ x 8½. 23435-5 Pa. $4.00

THE EARLY WORK OF AUBREY BEARDSLEY, Aubrey Beardsley. 157 plates, 2 in color: *Manon Lescaut, Madame Bovary, Morte Darthur, Salome,* other. Introduction by H. Marillier. 182pp. 8⅛ x 11. 21816-3 Pa. $4.50

THE LATER WORK OF AUBREY BEARDSLEY, Aubrey Beardsley. Exotic masterpieces of full maturity: *Venus and Tannhauser, Lysistrata, Rape of the Lock, Volpone,* Savoy material, etc. 174 plates, 2 in color. 186pp. 8⅛ x 11. 21817-1 Pa. $4.50

THOMAS NAST'S CHRISTMAS DRAWINGS, Thomas Nast. Almost all Christmas drawings by creator of image of Santa Claus as we know it, and one of America's foremost illustrators and political cartoonists. 66 illustrations. 3 illustrations in color on covers. 96pp. 8⅜ x 11¼. 23660-9 Pa. $3.50

THE DORÉ ILLUSTRATIONS FOR DANTE'S DIVINE COMEDY, Gustave Doré. All 135 plates from Inferno, Purgatory, Paradise; fantastic tortures, infernal landscapes, celestial wonders. Each plate with appropriate (translated) verses. 141pp. 9 x 12. 23231-X Pa. $4.50

DORÉ'S ILLUSTRATIONS FOR RABELAIS, Gustave Doré. 252 striking illustrations of *Gargantua and Pantagruel* books by foremost 19th-century illustrator. Including 60 plates, 192 delightful smaller illustrations. 153pp. 9 x 12. 23656-0 Pa. $5.00

LONDON: A PILGRIMAGE, Gustave Doré, Blanchard Jerrold. Squalor, riches, misery, beauty of mid-Victorian metropolis; 55 wonderful plates, 125 other illustrations, full social, cultural text by Jerrold. 191pp. of text. 9⅜ x 12¼. 22306-X Pa. $6.00

THE RIME OF THE ANCIENT MARINER, Gustave Doré, S. T. Coleridge. Dore's finest work, 34 plates capture moods, subtleties of poem. Full text. Introduction by Millicent Rose. 77pp. 9¼ x 12. 22305-1 Pa. $3.50

THE DORE BIBLE ILLUSTRATIONS, Gustave Doré. All wonderful, detailed plates: Adam and Eve, Flood, Babylon, Life of Jesus, etc. Brief King James text with each plate. Introduction by Millicent Rose. 241 plates. 241pp. 9 x 12. 23004-X Pa. $6.00

THE COMPLETE ENGRAVINGS, ETCHINGS AND DRYPOINTS OF ALBRECHT DURER. "Knight, Death and Devil"; "Melencolia," and more—all Dürer's known works in all three media, including 6 works formerly attributed to him. 120 plates. 235pp. 8⅜ x 11¼. 22851-7 Pa. $6.50

MAXIMILIAN'S TRIUMPHAL ARCH, Albrecht Dürer and others. Incredible monument of woodcut art: 8 foot high elaborate arch—heraldic figures, humans, battle scenes, fantastic elements—that you can assemble yourself. Printed on one side, layout for assembly. 143pp. 11 x 16. 21451-6 Pa. $5.00

UNCLE SILAS, J. Sheridan LeFanu. Victorian Gothic mystery novel, considered by many best of period, even better than Collins or Dickens. Wonderful psychological terror. Introduction by Frederick Shroyer. 436pp. 5⅜ x 8½. 21715-9 Pa. $6.00

JURGEN, James Branch Cabell. The great erotic fantasy of the 1920's that delighted thousands, shocked thousands more. Full final text, Lane edition with 13 plates by Frank Pape. 346pp. 5⅜ x 8½. 23507-6 Pa. $4.50

THE CLAVERINGS, Anthony Trollope. Major novel, chronicling aspects of British Victorian society, personalities. Reprint of Cornhill serialization, 16 plates by M. Edwards; first reprint of full text. Introduction by Norman Donaldson. 412pp. 5⅜ x 8½. 23464-9 Pa. $5.00

KEPT IN THE DARK, Anthony Trollope. Unusual short novel about Victorian morality and abnormal psychology by the great English author. Probably the first American publication. Frontispiece by Sir John Millais. 92pp. 6½ x 9¼. 23609-9 Pa. $2.50

RALPH THE HEIR, Anthony Trollope. Forgotten tale of illegitimacy, inheritance. Master novel of Trollope's later years. Victorian country estates, clubs, Parliament, fox hunting, world of fully realized characters. Reprint of 1871 edition. 12 illustrations by F. A. Faser. 434pp. of text. 5⅜ x 8½. 23642-0 Pa. $5.00

YEKL and THE IMPORTED BRIDEGROOM AND OTHER STORIES OF THE NEW YORK GHETTO, Abraham Cahan. Film *Hester Street* based on *Yekl* (1896). Novel, other stories among first about Jewish immigrants of N.Y.'s East Side. Highly praised by W. D. Howells—Cahan "a new star of realism." New introduction by Bernard G. Richards. 240pp. 5⅜ x 8½. 22427-9 Pa. $3.50

THE HIGH PLACE, James Branch Cabell. Great fantasy writer's enchanting comedy of disenchantment set in 18th-century France. Considered by some critics to be even better than his famous *Jurgen*. 10 illustrations and numerous vignettes by noted fantasy artist Frank C. Pape. 320pp. 5⅜ x 8½. 23670-6 Pa. $4.00

ALICE'S ADVENTURES UNDER GROUND, Lewis Carroll. Facsimile of ms. Carroll gave Alice Liddell in 1864. Different in many ways from final Alice. Handlettered, illustrated by Carroll. Introduction by Martin Gardner. 128pp. 5⅜ x 8½. 21482-6 Pa. $2.00

FAVORITE ANDREW LANG FAIRY TALE BOOKS IN MANY COLORS, Andrew Lang. The four Lang favorites in a boxed set—the complete *Red, Green, Yellow* and *Blue* Fairy Books. 164 stories; 439 illustrations by Lancelot Speed, Henry Ford and G. P. Jacomb Hood. Total of about 1500pp. 5⅜ x 8½. 23407-X Boxed set, Pa. $14.95

HOUSEHOLD STORIES BY THE BROTHERS GRIMM. All the great Grimm stories: "Rumpelstiltskin," "Snow White," "Hansel and Gretel," etc., with 114 illustrations by Walter Crane. 269pp. 5⅜ x 8½.
21080-4 Pa. $3.00

SLEEPING BEAUTY, illustrated by Arthur Rackham. Perhaps the fullest, most delightful version ever, told by C. S. Evans. Rackham's best work. 49 illustrations. 110pp. 7⅞ x 10¾. 22756-1 Pa. $2.50

AMERICAN FAIRY TALES, L. Frank Baum. Young cowboy lassoes Father Time; dummy in Mr. Floman's department store window comes to life; and 10 other fairy tales. 41 illustrations by N. P. Hall, Harry Kennedy, Ike Morgan, and Ralph Gardner. 209pp. 5⅜ x 8½. 23643-9 Pa. $3.00

THE WONDERFUL WIZARD OF OZ, L. Frank Baum. Facsimile in full color of America's finest children's classic. Introduction by Martin Gardner. 143 illustrations by W. W. Denslow. 267pp. 5⅜ x 8½.
20691-2 Pa. $3.50

THE TALE OF PETER RABBIT, Beatrix Potter. The inimitable Peter's terrifying adventure in Mr. McGregor's garden, with all 27 wonderful, full-color Potter illustrations. 55pp. 4¼ x 5½. (Available in U.S. only)
22827-4 Pa. $1.25

THE STORY OF KING ARTHUR AND HIS KNIGHTS, Howard Pyle. Finest children's version of life of King Arthur. 48 illustrations by Pyle. 131pp. 6⅛ x 9¼. 21445-1 Pa. $4.95

CARUSO'S CARICATURES, Enrico Caruso. Great tenor's remarkable caricatures of self, fellow musicians, composers, others. Toscanini, Puccini, Farrar, etc. Impish, cutting, insightful. 473 illustrations. Preface by M. Sisca. 217pp. 8⅜ x 11¼. 23528-9 Pa. $6.95

PERSONAL NARRATIVE OF A PILGRIMAGE TO ALMADINAH AND MECCAH, Richard Burton. Great travel classic by remarkably colorful personality. Burton, disguised as a Moroccan, visited sacred shrines of Islam, narrowly escaping death. Wonderful observations of Islamic life, customs, personalities. 47 illustrations. Total of 959pp. 5⅜ x 8½.
21217-3, 21218-1 Pa., Two-vol. set $12.00

INCIDENTS OF TRAVEL IN YUCATAN, John L. Stephens. Classic (1843) exploration of jungles of Yucatan, looking for evidences of Maya civilization. Travel adventures, Mexican and Indian culture, etc. Total of 669pp. 5⅜ x 8½. 20926-1, 20927-X Pa., Two-vol. set $7.90

AMERICAN LITERARY AUTOGRAPHS FROM WASHINGTON IRVING TO HENRY JAMES, Herbert Cahoon, et al. Letters, poems, manuscripts of Hawthorne, Thoreau, Twain, Alcott, Whitman, 67 other prominent American authors. Reproductions, full transcripts and commentary. Plus checklist of all American Literary Autographs in The Pierpont Morgan Library. Printed on exceptionally high-quality paper. 136 illustrations. 212pp. 9⅛ x 12¼. 23548-3 Pa. $7.95